Refining Chemical Processes For Environmental Stewardship

Sustainable Production's Pillars And Efficient Manufacturing Practices

Kaleb Abernathy

Table of Contents

1 Introduction to Green Chemistry **5**
- 1.1 Principles of Green Chemistry 8
- 1.2 Metrics for Sustainability 13

2 Process Intensification Techniques **17**
- 2.1 Microreactor Technology 20
- 2.2 Membrane Separations 23
- 2.3 Supercritical Fluid Extraction 26

3 Waste Minimization Strategies **31**
- 3.1 Waste Audits & Reduction 33
- 3.2 Recycling & Reuse Methods 36
- 3.3 Waste Treatment Technologies 40
- 3.4 Pollution Prevention 43
- 3.5 Clean Production Technologies 46

4 Sustainable Solvents & Catalysts **51**
- 4.1 Ionic Liquids 54
- 4.2 Bio-based Catalysts 57

5 Energy Efficiency in Chemical Processes **61**
- 5.1 Process Optimization 64
- 5.2 Heat Integration 67
- 5.3 Renewable Energy Sources 69

6 Life Cycle Assessment (LCA) **73**
6.1 LCA Methodology 76
6.2 Interpreting LCA Results 79

7 Case Studies & Future Directions **83**
7.1 Industrial Examples 86
7.2 Emerging Technologies 89
7.3 Policy & Regulation 93
7.4 Research Opportunities 95
7.5 Challenges & Opportunities 98

1 Introduction to Green Chemistry

Green chemistry isn't merely a set of guidelines; it's a fundamental shift in how we approach chemical processes. It's about designing chemical products and processes that minimize or eliminate the use and generation of hazardous substances. This proactive approach stands in stark contrast to traditional methods that often rely on end-of-pipe remediation to address pollution after it's already occurred. The core principle lies in preventing pollution at its source, creating a fundamentally safer and more sustainable chemical industry.

The twelve principles of green chemistry provide a framework for achieving this transformation. They encourage the use of renewable feedstocks, designing safer chemicals, minimizing waste, designing energy-efficient processes, and choosing safer solvents and reaction conditions, among others. These principles are not mutually exclusive; they often work in synergy to achieve a more holistic approach to sustainability. Consider, for instance, how designing a process with high atom economy inherently minimizes waste, aligning with multiple principles simultaneously. This integrated approach is crucial for comprehensive impact.

Implementing green chemistry principles requires a systematic approach. One cannot simply replace one chemical with another without considering the broader consequences. For instance, switching to a supposedly "green" solvent might inadvertently introduce other environmental or health concerns if not fully evaluated within the context of the entire process. A thorough life cycle assessment (LCA) is essential. This rigorous analysis assesses the environmental impact of a product or process throughout its entire lifecycle, from raw material extraction to end-of-life disposal.

This comprehensive evaluation provides a holistic perspective on the sustainability of the entire production process and its various stages. LCA also enables comparative analysis of different options, guiding informed decisions. In other words, understanding the impact of a chemical process in its totality is paramount, hence the critical role LCA plays in the transition to green chemistry. A well-conducted LCA can unearth hidden environmental impacts that might not be immediately apparent from a superficial examination.

Beyond LCA, several metrics offer valuable quantitative measures of a process's environmental performance. Atom economy, for example, calculates the efficiency of a chemical reaction by comparing the mass of desired product to the total mass of reactants. A higher atom economy indicates less waste generation. Similarly, the E-factor, which relates the mass of waste generated to the mass of the desired product, provides a direct measure of waste generation. Lower E-factors are synonymous with greener processes.

The choice of solvents and catalysts significantly impacts the environmental footprint of a chemical reaction. Traditional solvents often pose health and environmental hazards. The transition to greener alternatives, such as ionic liquids or supercritical fluids, can dramatically reduce these risks. Ionic liquids, for instance, offer unique properties that allow for more efficient and environmentally benign reactions. They can be designed with tailored properties, providing flexibility and control over reaction conditions. In the same vein, bio-based catalysts, derived from renewable resources, contribute to a circular economy.

Energy efficiency is another key aspect of green chemistry. Minimizing energy consumption reduces the carbon footprint and operational costs of chemical processes. Techniques like process intensification, using microreactors to enhance heat and mass transfer, and heat integration to recover and reuse waste heat are crucial steps towards energy efficiency. Integrating renewable energy sources, such as solar or wind power, further contributes to achieving sustainability goals. Implementing energy-saving measures might require upfront investment, however, the long-term benefits in terms of cost savings and environmental protection far outweigh the initial expenses.

The successful implementation of green chemistry requires a collaborative effort among researchers, engineers, policymakers, and industry stakeholders. Regulations and incentives play a pivotal role in driving the adoption of sustainable practices. Research continues to push the boundaries of green chemistry, developing new technologies and approaches to make chemical production even more environmentally friendly. The transition towards a truly sustainable chemical industry requires the concerted efforts of every player involved in the chemical landscape, from academics to industry leaders and governments. The future of chemical production depends on successfully embracing these principles, leading us towards a greener and more sustainable future for everyone.

1.1 Principles of Green Chemistry

The twelve principles of green chemistry provide a framework for designing and implementing environmentally benign chemical processes. Prevention of waste is paramount; it's far more efficient and sustainable to design processes that minimize waste generation from the outset rather than trying to clean up pollution after it's created. This proactive approach is far more cost-effective in the long run, reducing both financial and environmental burdens. A crucial aspect of this preventative strategy involves the design of chemical products to minimize toxicity. This involves considering the entire lifecycle of the product, from its synthesis to its eventual disposal, ensuring it doesn't pose a risk to human health or the environment at any stage. This holistic approach demands careful selection of starting materials and reaction pathways.

Atom economy, a key metric in green chemistry, focuses on maximizing the incorporation of all materials used in a chemical reaction into the final product. This minimizes waste, a direct implementation of the principle of waste prevention. High atom economy translates to reduced waste disposal costs and a smaller environmental footprint, enhancing the overall sustainability of the chemical process. The opposite, a low atom economy, typically results in a large amount of byproduct, leading to increased waste generation and disposal challenges. Optimizing processes for maximum atom economy is not merely a matter of reducing waste; it's also a way to improve the overall efficiency and cost-effectiveness of the chemical process. Implementing these principles leads to a fundamental shift from treating waste as an afterthought to integrating environmental considerations into the very core of chemical process design.

Designing safer chemicals, another central principle, goes beyond minimizing toxicity. It actively seeks to create chemical products and processes that are inherently less hazardous. This involves careful selection of less hazardous chemicals, or designing chemicals that naturally degrade after use. For instance, biodegradable materials are a classic example, mitigating the long-term persistence of harmful compounds in the environment. This principle actively pushes towards a circular economy, with a focus on the ultimate fate of all chemical materials in the environment. This requires a holistic and far-reaching approach to chemical synthesis and product design.

Energy efficiency is paramount for reducing the overall environmental impact. The creation of chemical products is an energy-intensive process, and minimizing energy consumption is essential for environmental responsibility. It means employing efficient reaction conditions, using innovative heat management systems, optimizing reaction pathways, and exploring renewable energy sources. By reducing energy consumption, we directly lessen reliance on fossil fuels, significantly reducing greenhouse gas emissions and our overall carbon footprint. Improving energy efficiency contributes to a more sustainable chemical industry, reducing costs and environmental harm simultaneously.

The use of renewable feedstocks is fundamental to sustainable chemistry. Replacing fossil fuel-based raw materials with renewable resources reduces our reliance on finite resources and lowers greenhouse gas emissions. Bio-based feedstocks, derived from plants or other renewable sources, are becoming increasingly significant in the chemical industry. Using renewable feedstocks also promotes a circular economy, transitioning from a linear "take-make-dispose" model to one where resources are used more efficiently and wastes are minimized. This circular approach drastically cuts the environmental impacts associated with extraction and processing of fossil fuels, promoting a more resilient and sustainable industrial ecosystem.

Catalysis is critical in green chemistry. Catalysts can significantly enhance the efficiency of chemical reactions, often leading to lower energy consumption and reduced waste generation. Furthermore, catalysts can often be used in smaller amounts, minimizing their overall environmental impact. Highly selective catalysts allow for cleaner reactions and reduce unwanted byproducts. This principle aims to design processes with high yields while minimizing waste generation and maximizing resource efficiency. The intelligent application of catalysis promotes efficiency and significantly decreases the need for excessive amounts of chemicals. It's an indispensable strategy for achieving sustainable chemical transformations. Reducing the use of chemical derivatives in synthesizing processes is a significant element of green chemistry. Derivatives, often added to modify a molecule's reactivity, can introduce additional steps and increase the amount of waste. By reducing the dependence on such derivatives, we can simplify reaction pathways, minimize waste, and enhance the overall efficiency of the chemical process. In many cases, designing reactions that avoid derivatives entirely is possible, contributing to a cleaner and more environmentally friendly production method. This is a specific approach to green chemistry that directly influences the efficiency and environmental impact of the manufacturing processes.

Designing chemicals and processes for degradation ensures that chemical products break down into innocuous substances after their use, preventing the accumulation of persistent pollutants in the environment. This principle aims to reduce the overall environmental impact by limiting the longevity and potential harm of products released into the environment. Biodegradable materials are an important example. Carefully designed chemicals can decompose safely, thus avoiding the accumulation of harmful substances in the environment, contributing significantly to environmental protection.

Real-time analysis for pollution prevention is a powerful tool in green chemistry. Monitoring and analyzing chemical processes in real-time allows for immediate adjustments and problem-solving, preventing the generation of pollution in the first place. It's a proactive approach that significantly reduces the environmental impact. This includes implementing continuous monitoring systems and automated controls, allowing for quick responses to deviations and preventing major pollution events. This shifts focus away from the reactive cleanup of pollution to its preventative control.

Minimizing the potential for accidents is vital in chemical production. Processes should be inherently safer, minimizing the risk of explosions, fires, or releases of hazardous substances. This involves careful selection of materials, robust process design, and appropriate safety protocols. Implementing robust safety measures prevents major environmental catastrophes and protects human health. This approach to process design ensures a safe working environment and minimizes potential damage from accidents. Safe processes ultimately benefit both people and the environment. The design of inherently safer chemistry focuses on using substances that pose minimal risk. This involves evaluating materials based on their toxicity and environmental persistence, and prioritizing the use of safer alternatives. This proactive approach shifts the focus from mitigating the risks of hazardous materials to employing only safer substances from the outset. Adopting this strategy inherently protects both the environment and human health, improving the overall safety and sustainability of the chemical process.

Finally, the concept of designing for the end-of-life of materials is crucial for environmentally benign chemistry. Considering the end-of-life of a chemical product from its initial design helps create products that are easier to recycle, reuse, or dispose of safely, minimizing environmental damage. This strategy encourages a circular economy model, promoting efficient resource use and waste reduction. It's a holistic approach, addressing the environmental impact across the entire lifecycle of a chemical product, creating a more responsible and sustainable chemical industry.

1.2 Metrics for Sustainability

Selecting appropriate metrics is crucial for effectively evaluating the environmental performance of chemical processes. We need quantifiable measures to track progress, identify areas for improvement, and demonstrate the impact of sustainability initiatives. Simply stating a process is "greener" isn't sufficient; robust metrics provide concrete evidence. This chapter delves into several key metrics, explaining their calculation and application.

Atom economy, a fundamental concept in green chemistry, directly assesses the efficiency of a chemical reaction. It measures the percentage of starting materials incorporated into the desired product. A higher atom economy indicates less waste generation, a key tenet of sustainable production. Calculations are straightforward, requiring only the molecular weights of reactants and products. For instance, a reaction with a 100% atom economy means all starting materials are converted into the final product, minimizing waste. However, it's important to note atom economy primarily focuses on the reaction itself, not the overall process.

The E-factor, or Environmental Factor, provides a broader perspective by considering the mass of waste generated relative to the mass of the desired product. Calculated as the ratio of waste to product mass, a lower E-factor signifies better environmental performance. For instance, pharmaceutical manufacturing often faces significantly higher E-factors than bulk chemical production due to complex reaction sequences and extensive purification steps. Analyzing the E-factor helps pinpoint process steps generating the most waste, informing targeted improvements. This metric reveals the overall environmental burden of a process, encompassing both reaction and downstream activities.

Beyond reaction efficiency, the process's energy consumption significantly impacts its environmental footprint. Energy intensity, the amount of energy required per unit of product, acts as a critical metric. It's essential to consider both direct energy use (heating, stirring, etc.) and indirect energy consumption (manufacturing equipment, transportation of materials). Reducing energy intensity lowers greenhouse gas emissions and operational costs, demonstrating a dual benefit. This involves scrutinizing energy usage in each step to pinpoint opportunities for improvement. Efficient energy management systems coupled with renewable energy integration play a pivotal role.

Life Cycle Assessment (LCA), a more comprehensive approach discussed later in the book, examines the environmental impact of a chemical product throughout its entire lifecycle, from raw material extraction to disposal. While complex, LCA offers a holistic view, considering factors often missed by simpler metrics. It encompasses energy use, emissions, water consumption, and waste generation, providing a more complete picture of the process's environmental impact. Remember, LCA requires comprehensive data acquisition and sophisticated analysis, making it more resource-intensive compared to other metrics.

While these individual metrics offer valuable insights, integrating them provides the most comprehensive evaluation. Combining atom economy with the E-factor and energy intensity creates a more nuanced understanding of the process's sustainability. For example, a high atom economy reaction might still yield a high E-factor if substantial solvent usage or purification steps contribute to significant waste. Therefore, a balanced assessment incorporates multiple metrics, offering a more thorough appraisal of environmental performance. This holistic evaluation directs efficient resource allocation towards optimizing processes for improved sustainability.

Furthermore, don't underestimate the importance of context. The relevance of each metric can vary depending on the specific chemical process, its scale, and the environmental priorities of the company or industry. What's most important for one process might not be as critical for another. Contextual analysis allows for a customized strategy in improving sustainability, adapting to the specific demands of different processes. A flexible, adaptable approach is necessary for practical application and successful implementation.

Finally, remember that these metrics are tools, not ends in themselves. The ultimate goal is to continuously improve environmental performance. Regularly monitoring and analyzing these metrics enables ongoing improvements. Sustainable chemical production is a journey, not a destination; embracing this ethos is fundamental to long-term success. These indicators offer guidance along this path, encouraging a proactive approach to environmental stewardship. Continuous monitoring and refinement are vital for consistent progress in environmental sustainability.

2 Process Intensification Techniques

Microreactors, for instance, offer significantly enhanced heat and mass transfer compared to conventional reactors. Their small scale allows for precise control over reaction parameters, leading to improved yields and selectivity. This translates directly into reduced waste generation and a more efficient use of resources, key aspects of sustainable chemical production. Imagine scaling down a reaction from a large vat to a network of tiny channels—the implications for safety and efficiency are dramatic.

Membrane separations present another powerful tool in the process intensification arsenal. These techniques exploit selective permeability to separate different components of a mixture, often with minimal energy consumption. Unlike traditional distillation or extraction methods, membrane-based separations can offer significantly lower operating costs and reduced environmental impact, especially when dealing with complex mixtures or sensitive compounds. Consider the potential for purifying valuable products and recycling solvents using innovative membrane technology; the reduction in waste disposal costs alone can be substantial.

Supercritical fluid extraction (SFE) represents a further advancement in environmentally friendly separations. Supercritical fluids, such as carbon dioxide, possess unique properties that allow for efficient extraction of target compounds from various matrices. The process typically operates at moderate temperatures and pressures, minimizing energy consumption and eliminating the need for harsh organic solvents. This method has shown great promise in extracting valuable components from natural products, thus offering a path towards more sustainable supply chains. The reduced toxicity and ease of recovery make it especially attractive for a wide range of applications.

The combination of these process intensification techniques often yields synergistic effects. Imagine integrating microreactors with membrane separations to achieve a truly continuous and efficient process flow. Such integrated systems can drastically reduce waste, improve product quality, and enhance overall safety. The ability to automate and precisely control every step of the manufacturing process, from synthesis to purification, is transformative in terms of both economic viability and environmental stewardship.

Beyond these specific techniques, the overarching principle of process intensification lies in its focus on minimizing the overall footprint of chemical production. By integrating multiple unit operations into a single, compact system, process intensification reduces the need for extensive infrastructure, simplifies process control, and significantly minimizes waste. This approach directly supports a more efficient and environmentally sound manufacturing process. Consider the long-term benefits of reduced capital costs, lower operational expenses, and a smaller environmental impact. This approach isn't merely about efficiency; it's a paradigm shift in how we design and operate chemical processes, moving toward a truly sustainable future.

Adopting process intensification technologies often requires an initial investment in new equipment and expertise. However, the long-term benefits—reduced waste disposal costs, improved product yields, and lower energy consumption—significantly outweigh these initial expenses. Moreover, the potential for increased profitability, driven by reduced operating costs and improved product quality, makes these technologies increasingly attractive to a wider range of chemical manufacturers. Remember, investing in sustainability is an investment in the future. The long-term gains in economic and environmental stability are immense.

Careful consideration of process design, material selection, and operational parameters is crucial for the successful implementation of process intensification strategies. The choice of appropriate materials must balance performance requirements with sustainability goals, favoring readily recyclable and less toxic options whenever possible. Likewise, process optimization strategies should aim to minimize energy consumption, waste generation, and the release of harmful substances. The choice of material is critical. A holistic approach is essential for success.

The transition to intensified processes can be facilitated by a multi-faceted approach. This includes collaboration between academia and industry to develop and implement new technologies, robust regulatory frameworks that incentivize the adoption of sustainable practices, and targeted training programs to equip the workforce with the necessary skills. A concerted effort from all stakeholders is vital. Only through collaboration can we overcome the initial barriers to widespread adoption and achieve the transformative potential of process intensification. This strategy represents a commitment to a more sustainable chemical industry.

Finally, embracing process intensification is not simply a matter of technological advancement; it is a cultural shift. It requires a paradigm shift in how we approach chemical process design, emphasizing efficiency, sustainability, and safety from the outset. By designing inherently safer processes and integrating environmentally benign technologies, we can move beyond traditional end-of-pipe solutions to a more holistic approach to pollution prevention. A cultural shift is as crucial as the technology itself. This cultural change represents a fundamental commitment to environmental responsibility. It's about reimagining what is possible.

2.1 Microreactor Technology

Microreactors offer a compelling solution for enhancing reaction control and efficiency within chemical processes. Their small size, typically ranging from millimeters to centimeters, enables precise manipulation of reaction parameters, leading to superior product yields and reduced waste generation. This precise control stems from the significantly increased surface area-to-volume ratio inherent in microreactor design, allowing for rapid heat and mass transfer. Consequently, reactions proceed more quickly and uniformly, minimizing the potential for side reactions and improving overall selectivity. This results in higher purity products and less need for downstream purification, streamlining the production process and reducing the environmental impact.

The unique characteristics of microreactors also translate to improved safety. Reactions involving hazardous materials or exothermic processes become inherently safer due to the smaller reaction volumes. Should a runaway reaction occur, the limited scale confines the incident, mitigating the risks associated with larger-scale reactors. This enhanced safety profile contributes significantly to a more sustainable and responsible production environment, protecting both workers and the surrounding ecosystem. Furthermore, the ability to conduct a variety of reaction types within these small devices—from simple mixing to more complex multi-step processes—makes microreactors exceptionally versatile tools in chemical synthesis.

Designing and implementing a microreactor system requires careful consideration of several key factors. Selecting the appropriate material for the microchannels is crucial. The chosen material must be compatible with the specific reactants and reaction conditions, ensuring chemical inertness and resistance to corrosion. Additionally, the channel dimensions need to be optimized to promote efficient mixing and heat transfer, which is critically dependent on the fluid dynamics within the system. Careful simulation and modeling, often employing computational fluid dynamics (CFD), are essential to achieving optimal performance. This design phase must balance the advantages of miniaturization with practical considerations such as ease of manufacturing, scalability, and maintenance.

Beyond the reactor itself, the overall system design is also vital. Precise control over flow rates, temperature, and pressure is essential to maintain consistent reaction conditions. This often involves integrating specialized pumps, heaters, and sensors to monitor and regulate these parameters in real-time. Moreover, the integration of downstream processing units, such as separators or purifiers, directly into the microreactor system can further improve efficiency and reduce waste. This creates a compact and efficient production line, drastically reducing the footprint of the chemical process. Advanced control systems, using sophisticated algorithms and real-time data analysis, are frequently employed to optimize the entire system's performance, maximizing yield and minimizing energy consumption.

The scalability of microreactor technology is an important consideration for widespread adoption. While individual microreactors are relatively small, they can be easily integrated in parallel arrays or networks to achieve significant production capacity. This modular design allows for flexible scaling up, adapting the production level to match the demand without compromising the benefits of microreactor technology. This modularity is particularly attractive to chemical manufacturers, as it enables a gradual transition to sustainable processes while minimizing disruptions to existing operations. Furthermore, the reduced volume of reactants and solvents needed in microreactors reduces waste generation, resulting in significant cost savings on materials and waste disposal. This economic advantage complements the environmental benefits, making microreactors an attractive proposition for both large-scale and specialized chemical productions. The integration of microreactor technology into chemical processes is driving innovation across several industries, from pharmaceuticals and fine chemicals to energy and materials science. Its potential to drastically reduce waste, improve safety, and enhance efficiency is pushing researchers and engineers to explore new applications and optimize existing designs. The ongoing development of new materials, advanced fabrication techniques, and sophisticated control systems promises to expand the applicability of microreactors even further. Further research is needed into sustainable material selection for microreactor construction, exploration of novel geometries, and improved integration with other green chemistry principles, positioning microreactors as a pivotal technology for a more sustainable future of chemical manufacturing. The resulting advancements in reactor design and process optimization will not only bolster the competitiveness of environmentally responsible manufacturing processes, but they will also play a decisive role in shaping a more sustainable and environmentally conscious chemical industry.

2.2 Membrane Separations

Membrane separations offer a compelling pathway towards greener chemical processes. These technologies, encompassing a range of techniques like microfiltration, ultrafiltration, nanofiltration, reverse osmosis, and electrodialysis, excel at separating mixtures based on size, charge, or other physicochemical properties. Their effectiveness stems from the use of selectively permeable membranes, acting as efficient barriers that minimize energy consumption and reduce the need for harsh chemicals often associated with traditional separation methods. This makes them particularly valuable in minimizing waste streams and enhancing overall process efficiency, key tenets of environmentally responsible manufacturing.

Consider the considerable advantages of membrane separation over conventional techniques like distillation. Distillation, while widely used, often demands significant energy input, leading to substantial greenhouse gas emissions. Membranes, conversely, operate at milder temperatures and pressures, requiring less energy. This reduction in energy consumption translates directly into lower operating costs and a smaller carbon footprint, making them economically and environmentally advantageous. Furthermore, membrane technology often avoids the use of volatile organic compounds (VOCs) and other hazardous solvents, eliminating a major source of pollution inherent in several traditional processes.

Choosing the right membrane depends heavily on the specific separation challenge. The selection process requires a careful assessment of several factors, including the nature of the feed stream, the desired degree of separation, and the operating conditions. For instance, microfiltration might suit removing suspended solids from a liquid, while reverse osmosis excels at desalinating water or concentrating solutions. Electrodialysis effectively separates ionic species, proving particularly useful in various wastewater treatment applications. Proper membrane selection is crucial for optimizing both separation performance and overall sustainability goals.

The integration of membrane separations into a chemical process can often lead to a more compact and efficient plant layout. Replacing multiple unit operations with a single membrane-based system streamlines the process, reduces capital investment, and minimizes the overall footprint of the manufacturing facility. This spatial efficiency extends beyond mere footprint reduction; it also minimizes the potential for leaks and spills, further enhancing safety and reducing environmental risks. A streamlined process inherently translates to easier monitoring and control, thus enhancing the predictability and reliability of the output.

However, it's crucial to recognize the limitations of membrane separations. Membrane fouling, where accumulated substances clog the pores and reduce efficiency, remains a significant challenge. Regular cleaning or replacement is often required, impacting both operating costs and the overall environmental profile. The selection of appropriate pre-treatment steps to minimize fouling is therefore critical, as is the careful choice of membrane materials with enhanced resistance to fouling. Research into novel, antifouling membrane materials represents a promising avenue for mitigating this issue and improving the long-term sustainability of this separation method. Understanding these limitations and addressing them through innovative approaches is key to maximizing the full potential of membrane technology in advancing sustainable chemical production.

Beyond the immediate benefits of reduced energy consumption and waste generation, incorporating membrane separations contributes to the broader goal of a circular economy. By efficiently separating valuable components from waste streams, membranes enable recycling and reuse, conserving valuable resources and minimizing reliance on virgin materials. This contributes to reduced reliance on landfills and minimizes the depletion of finite natural resources—an invaluable contribution to long-term environmental stewardship. The integration of membrane technology supports a closed-loop system, wherein waste is viewed not as a disposal problem but as a potential source of valuable resources. This paradigm shift significantly impacts the environmental footprint of the chemical manufacturing industry. Ultimately, embracing membrane technologies in chemical processes represents a step towards a more sustainable and environmentally responsible future for the industry.

2.3 Supercritical Fluid Extraction

Supercritical fluid extraction (SFE) offers a powerful, environmentally friendly alternative to traditional solvent extraction methods. Its unique properties stem from the supercritical state of a substance, existing beyond its critical temperature and pressure. In this state, the solvent exhibits properties of both a liquid and a gas, boasting high solvating power like a liquid, yet possessing the low viscosity and high diffusivity of a gas. This combination allows for efficient extraction of a wide range of compounds from diverse matrices, minimizing the use of harmful organic solvents.

The process itself is relatively straightforward. The chosen supercritical fluid, typically carbon dioxide (CO2) due to its low critical temperature (31.1 °C) and pressure (73.8 bar), is pumped into an extraction vessel containing the material of interest. Precise control of temperature and pressure allows for selective extraction of specific components. After extraction, the pressure is carefully reduced, causing the supercritical fluid to revert to its gaseous state, leaving behind the extracted compounds. This simple process significantly reduces waste generation and avoids the need for energy-intensive post-processing steps often associated with conventional methods. CO2, being non-toxic and readily available, further enhances the method's green credentials.

One significant advantage of SFE lies in its exceptional selectivity. By carefully adjusting operating parameters, such as temperature and pressure, one can target specific compounds or fractions within a complex mixture. This precision minimizes unwanted co-extraction and simplifies downstream processing, leading to higher purity products and reduced waste. For instance, SFE has demonstrated success in extracting valuable components from natural products, like essential oils from herbs or active compounds from medicinal plants, without compromising their quality. Unlike conventional methods that might employ harsh organic solvents, potentially altering the composition or stability of the extracts, SFE maintains the integrity of the target compounds.

Furthermore, SFE's efficiency extends beyond its inherent selectivity. The high diffusivity of the supercritical fluid facilitates rapid mass transfer, shortening extraction times and increasing throughput. This improved efficiency translates into lower operational costs and a smaller environmental footprint compared to conventional methods. The process can be scaled up or down relatively easily depending on the need, from laboratory-scale extractions to large-scale industrial applications. Consider, for example, the extraction of caffeine from coffee beans; SFE offers a cleaner and more efficient alternative to traditional solvent extraction, producing decaffeinated coffee with minimal environmental impact. Such applications underscore SFE's potential to revolutionize various industries.

However, SFE isn't without its limitations. While CO2 is commonly used, its solvating power may be insufficient for certain polar compounds. Modifying the supercritical fluid with co-solvents can overcome this limitation but introduces additional complexities and may compromise the process's environmental friendliness. Cost can also be a factor, requiring specialized equipment and skilled operators. The initial investment may be high, although the long-term operational efficiency often offsets this cost. Careful consideration of the specific application and a thorough cost-benefit analysis are crucial before implementing SFE. Despite these limitations, the advantages of SFE often outweigh the disadvantages, particularly in applications where environmental stewardship is paramount.

Looking ahead, ongoing research is focused on expanding the applications of SFE and addressing some of its challenges. Developments in equipment design, coupled with a deeper understanding of the intricate thermodynamics and fluid dynamics involved, continue to improve the efficiency and effectiveness of the technique. Novel supercritical fluids are also being explored, broadening the range of extractable compounds. This continuous innovation ensures SFE remains a highly relevant and promising technology in the quest for more sustainable and environmentally responsible chemical processes. The future of SFE appears bright, promising even more environmentally friendly and efficient extraction capabilities in diverse industries. Its potential to improve sustainability makes SFE a crucial tool in the broader movement towards green chemistry. The exploration of supercritical fluids is constantly expanding and improving, making it a leading example of sustainable technology in various sectors. The ongoing research in this field constantly pushes the boundaries of what is possible, shaping a more environmentally conscious future for chemical processing.

3 Waste Minimization Strategies

Waste audits are the cornerstone of any effective waste minimization strategy. A thorough audit meticulously identifies all waste streams generated within a chemical process, quantifying their volume, composition, and hazardous characteristics. This detailed analysis reveals potential areas for improvement, pinpointing specific sources of waste generation and their associated environmental impacts. Effective auditing necessitates a collaborative approach, involving personnel from various departments to obtain a holistic view of the process. The collected data forms the basis for developing targeted waste reduction strategies, guiding investment decisions in waste treatment and recycling technologies and facilitating informed compliance with environmental regulations. Without this crucial initial step, efforts to minimize waste remain largely reactive and often inefficient.

Implementing effective recycling and reuse programs requires careful consideration of material properties and process compatibility. The feasibility of recycling often hinges on the purity and concentration of the waste stream. Pre-treatment processes, such as separation, purification, or concentration, may be necessary to prepare the waste for reuse or recycling.

Economic viability is paramount; the costs associated with pre-treatment, recycling, or reuse must be carefully weighed against the potential benefits, including reduced waste disposal costs and the recovery of valuable materials. A robust recycling program demands careful management, from proper storage and handling of recyclable materials to selecting appropriate recycling technologies, minimizing further environmental impact. These practices can lead to significant cost savings and a substantial reduction in waste disposal volumes.

Various waste treatment technologies exist, each with its own advantages and limitations. Incineration, for instance, offers a high degree of waste reduction but can generate air emissions that require stringent control. Landfilling, while relatively inexpensive, raises concerns about soil and groundwater contamination. Biological treatment methods, such as composting or anaerobic digestion, are environmentally friendly but may be suitable only for specific types of waste. The selection of an appropriate waste treatment technology necessitates a thorough assessment of the waste characteristics, environmental regulations, and economic considerations. Advanced treatment technologies, like plasma gasification or supercritical water oxidation, offer potentially more sustainable options but often come with higher capital and operational costs. The choice hinges upon a careful evaluation of all factors.

Pollution prevention, a proactive approach, focuses on eliminating waste generation at its source, rather than addressing it after the fact. This may involve process modifications, such as adopting cleaner production techniques, substituting hazardous materials with less harmful alternatives, or improving process efficiency to minimize waste formation. Implementing stricter quality control measures can also minimize waste generation by reducing the production of defective products. Adopting a pollution prevention philosophy requires a shift in mindset, from a reactive, end-of-pipe approach to a proactive, integrated approach across the entire production lifecycle. This necessitates comprehensive training of personnel and active participation from all levels of the organization.

Clean production technologies encompass a broad spectrum of innovative approaches aimed at minimizing waste generation and improving environmental performance. This includes technologies like process intensification, which minimizes equipment size and energy consumption while increasing throughput. It also covers the implementation of closed-loop systems to recycle or reuse process streams, reducing waste generation and improving resource efficiency. The adoption of these technologies can significantly reduce environmental impact and enhance economic viability. However, implementation requires a detailed assessment of the specific process, careful evaluation of costs and benefits, and a commitment to continuous improvement. Continuous monitoring and optimization are essential to ensure the long-term effectiveness of these technologies.

3.1 Waste Audits & Reduction

Conducting a thorough waste audit is the cornerstone of any effective waste reduction strategy. It involves systematically identifying and quantifying all waste streams generated within a chemical process. This detailed analysis goes beyond simply weighing the waste; it necessitates a meticulous examination of each waste's composition, its origin within the process, and its potential for reduction, reuse, or recycling. Remember, the goal isn't just to measure the waste; it's to understand it. This understanding provides the crucial foundation for making informed decisions about waste minimization. Detailed record-keeping, including flow diagrams and material balances, is paramount. This level of precision allows for accurate assessment of waste generation rates, enabling the identification of major contributors and providing a baseline for tracking progress in subsequent waste reduction efforts. Without this accurate data, targeted interventions become significantly more difficult. Accurate data, gathered systematically and consistently, allows for a data-driven approach that is essential in identifying the most effective solutions.

Data collection methods should be tailored to the specific process. This might involve direct observation, sampling analysis, review of existing process documentation or interviews with personnel involved in the process. Combine multiple methods for greater accuracy and a more comprehensive understanding of the waste streams. Consider using advanced analytical techniques, such as chromatography or spectroscopy, for a more detailed characterization of waste composition. The more you know about your waste streams, the more effectively you can target your waste reduction efforts, ensuring the best return on investment for your efforts. Remember to involve personnel from all levels within the operation. Their on-the-ground knowledge and experience are invaluable in identifying overlooked waste sources and suggesting efficient solutions.

Once your audit is complete and the data analyzed, prioritize waste reduction opportunities. Focus on the waste streams that contribute the most to the overall waste generation. Start with the 'low-hanging fruit.' These are usually the easiest and quickest to address, offering immediate reductions and building momentum for more ambitious projects. For example, implementing better housekeeping practices might immediately reduce packaging waste or spills. These low-cost solutions often yield significant environmental and economic benefits. After addressing the simpler areas, tackle more complex issues. These might involve redesigning the process itself or investigating alternative chemicals or technologies. Consider the total environmental impact of each action. This consideration is important for ensuring that proposed solutions don't simply shift the environmental burden to another stage of the process or to a different environmental medium.

Implementing the selected waste reduction strategies requires careful planning and execution. Develop a clear implementation plan that outlines each step, timelines, responsibilities, and resource allocation. This plan should be integrated into the broader operational strategy and should adhere to relevant safety regulations and best practices. Regular monitoring and evaluation are critical to track progress and make necessary adjustments. Use the data collected during the audit as a baseline to measure your progress. Tracking key performance indicators (KPIs) will not only inform of your progress but also support continuous improvement. A well-documented process of continuous improvement, including regular review and update of the waste reduction plan, will be crucial for the long-term success of your program.

Beyond the immediate improvements, consider the larger context of sustainability. A reduction in waste is not just about environmental protection; it translates into economic savings. Less waste means lower disposal costs and potentially the recovery of valuable materials. Furthermore, a reputation for environmental stewardship can enhance a company's image and attract customers and investors. These benefits extend beyond the immediate environmental gain. The entire process is about demonstrating a commitment to sustainability—both environmentally and economically. This holistic approach shows a commitment to responsible chemical manufacturing and generates a positive impact both on the company's bottom line and the environment. Remember that the information gathered during the waste audit and subsequent improvements provide valuable data for Life Cycle Assessments (LCA). This data can help in assessing the overall environmental impact of the process and inform future improvements. This reinforces the long-term value of a thorough and comprehensive waste reduction program.

3.2 Recycling & Reuse Methods

Effective recycling and reuse strategies are paramount for minimizing waste and maximizing resource utilization in chemical manufacturing. A thorough waste audit, as discussed in the previous chapter, identifies the types and quantities of waste generated, paving the way for targeted recycling and reuse initiatives. This detailed understanding forms the foundation for developing efficient and economically viable solutions. Without this crucial first step, any subsequent efforts risk being inefficient and ineffective.

Implementing a robust recycling program requires careful planning and execution. This includes establishing clear procedures for waste segregation and collection, ensuring that materials are properly sorted and handled to maintain their quality and prevent contamination. Proper labeling and documentation are equally vital for efficient tracking and management. For instance, separating solvents based on their chemical properties and potential for reuse is essential. This minimizes risks associated with incompatibility, guaranteeing the integrity of the recovered materials.

The selection of appropriate recycling technologies depends heavily on the nature of the waste. For example, physical methods such as filtration, sedimentation, and evaporation are suitable for separating mixtures of different phases. Chemical methods, including precipitation, oxidation, and reduction, may be employed to convert waste materials into more valuable or less harmful substances. In some cases, a combination of physical and chemical methods proves most effective in achieving the desired level of purification and recovery.

Reuse offers an even more sustainable alternative, minimizing the need for additional processing. For instance, spent solvents, after appropriate purification, might be reused directly in the same process or in a different, compatible application. This approach significantly reduces the need for fresh solvent procurement and minimizes the environmental burden associated with solvent disposal. Moreover, the reuse of process water, after appropriate treatment, can achieve remarkable water conservation and reduce overall environmental impact.

However, the economic viability of both recycling and reuse initiatives is crucial. The costs associated with waste collection, processing, and storage need to be carefully weighed against the potential benefits of material recovery and reduced disposal fees. Conducting a life-cycle cost analysis can provide valuable insight into the long-term economic feasibility of these strategies, guiding decision-making processes towards the most cost-effective and environmentally sound solutions.

Successful implementation requires effective training and engagement of personnel. Employees need to understand the importance of waste segregation, proper handling procedures, and the overall goals of the recycling and reuse program. Providing comprehensive training and clear guidelines empowers workers to actively contribute to waste reduction and environmental stewardship. Regular monitoring and evaluation are essential to identify areas for improvement and to adapt the program to evolving needs. Furthermore, collaboration with external partners, such as waste management companies and recycling specialists, can expand capabilities and access to advanced technologies. These partnerships can offer access to expertise and resources that might otherwise be unavailable, facilitating the implementation of complex or specialized recycling and reuse systems. This external expertise significantly enhances the effectiveness and efficiency of waste management initiatives.

Beyond technical considerations, regulatory compliance is paramount. Chemical manufacturers must adhere to relevant environmental regulations governing waste generation, handling, and disposal. Staying updated on evolving regulations and implementing best practices ensures compliance and prevents potential penalties. Regular audits and environmental impact assessments further strengthen compliance and demonstrate a commitment to environmental responsibility.

Technological advancements continue to drive innovation in recycling and reuse methods. Emerging technologies, such as advanced separation techniques and catalytic conversion processes, offer enhanced capabilities for waste valorization. Staying abreast of these innovations is critical to maximizing resource recovery and minimizing environmental impact. Embracing new technologies allows companies to achieve greater sustainability and maintain a competitive edge in a rapidly evolving market.

Ultimately, optimizing recycling and reuse strategies within chemical manufacturing demands a holistic and integrated approach. Careful planning, technological innovation, robust management systems, and consistent commitment to best practices are essential for success. Continuous improvement and adaptation are vital to maintaining a sustainable and efficient waste management program that minimizes environmental impact and optimizes resource utilization. This proactive approach not only reduces environmental burdens, but also fosters cost savings and enhances the overall sustainability profile of the company. By prioritizing recycling and reuse, chemical manufacturers demonstrate a deep commitment to environmental stewardship, and ultimately contribute towards a more sustainable future.

3.3 Waste Treatment Technologies

Effective waste treatment is paramount for environmentally responsible chemical production. Ignoring waste generation leads to environmental damage and regulatory issues, potentially crippling your business. Proactive waste management, however, demonstrates environmental stewardship, improves your public image, and might even unlock cost-saving opportunities through resource recovery.

Several technologies exist for treating chemical waste, each with strengths and weaknesses depending on the waste's composition and volume. Incineration, for example, effectively reduces waste volume but may produce air emissions requiring rigorous control, necessitating advanced air pollution control systems such as scrubbers and filters. The selection of the most appropriate technology requires a careful assessment of factors such as cost, efficiency, and the potential environmental impact of any byproducts. Remember to consult your local environmental agencies for guidance on acceptable practices.

Biological treatment, a more eco-friendly option, uses microorganisms to break down organic waste. This method is effective for biodegradable materials and can result in less hazardous byproducts compared to incineration, but its effectiveness depends heavily on the nature of the waste. Certain toxic chemicals may not be easily biodegradable, rendering this method unsuitable. For instance, halogenated organic compounds might require pretreatment or a combination of methods for effective treatment. Furthermore, the time required for biological treatment can be significantly longer than other methods.

Physical treatment methods, such as filtration, sedimentation, and evaporation, often serve as pretreatment steps before other methods are applied. These processes can remove solids, separate liquids, or concentrate waste streams, simplifying subsequent treatment. However, these methods often only transfer the waste problem, not solve it, needing to be combined with more advanced techniques. For example, filtering out solid waste from a liquid stream merely concentrates the contaminants, not destroy them, needing further treatment.

Chemical treatment neutralizes or alters the chemical properties of the waste to make it less hazardous. Methods like oxidation, reduction, and precipitation can transform toxic substances into less harmful forms. However, choosing the right chemical treatment requires expertise in chemistry and chemical engineering. Incorrectly implementing this process could lead to unexpected byproducts. Also, careful disposal or further treatment of the chemicals used in the process itself might be necessary.

Advanced oxidation processes (AOPs) represent a promising area. These processes utilize powerful oxidizing agents like ozone, hydrogen peroxide, or UV radiation to degrade even persistent organic pollutants. AOPs can be very effective, however, they often require significant energy input and can be expensive to implement on a large scale. They may also generate byproducts requiring further treatment or disposal. Careful consideration of operational costs and environmental consequences is crucial.

The choice of waste treatment technology is not a one-size-fits-all solution. A tailored approach based on a comprehensive waste characterization and a thorough life cycle assessment (LCA) is needed. This assessment will consider not only the treatment method itself but also the energy consumption, material usage, and overall environmental impact of all stages of the waste management process, from generation to disposal. Remember to consider the potential environmental benefits of adopting the best available technologies and practices.

The most sustainable approach, therefore, prioritizes waste minimization at its source. By implementing efficient process design, optimizing reaction conditions, and maximizing resource utilization, you can reduce waste generation from the outset, minimizing the need for extensive treatment. This proactive strategy often proves far more cost-effective and environmentally friendly than dealing with waste after it's generated. Prioritizing waste reduction before treatment is often the most economical and sustainable way to minimize environmental impact.

In conclusion, successfully managing waste from chemical processes necessitates careful consideration of various technologies and their respective merits. While various treatment methods exist, each has associated costs and environmental impacts. The most sustainable approach prioritizes waste minimization, then implements a combination of appropriate technologies selected after a comprehensive analysis. A holistic approach that considers factors such as energy consumption, and overall environmental impact is crucial to achieve sustainable and environmentally sound waste management practices. By combining an understanding of these factors with proactive waste reduction strategies, chemical producers can significantly reduce their environmental footprint and uphold their commitment to environmental stewardship.

3.4 Pollution Prevention

Pollution prevention isn't just about cleaning up messes; it's about fundamentally redesigning processes to eliminate waste at the source. This proactive approach, crucial for sustainable manufacturing, involves a shift in mindset, demanding a thorough understanding of your process and its inherent vulnerabilities. It's about asking not "how can we treat this waste?", but rather "how can we avoid generating it in the first place?" This requires detailed process analysis, identifying potential pollution hotspots, and implementing strategies to minimize or eliminate them.

Implementing effective pollution prevention hinges on a robust waste audit. This systematic evaluation, a crucial first step, identifies the types, quantities, and sources of waste generated throughout your process. Detailed record-keeping is essential, tracing waste streams back to their origins. This granular level of analysis facilitates the identification of specific process modifications that will minimize or eliminate waste at its source. Such audits often reveal surprising inefficiencies and opportunities for improvement, leading to significant cost savings alongside environmental benefits. Investing time and resources in conducting thorough waste audits pays off handsomely.

Process modifications form the core of pollution prevention. Consider substituting hazardous materials with less harmful alternatives. Bio-based solvents, for instance, often provide comparable performance with reduced environmental impact. Similarly, exploring alternative reaction pathways or employing catalytic processes can significantly reduce waste generation, a concept extensively explored in earlier chapters on sustainable solvents and catalysts. These modifications may require initial investment, but long-term benefits – reduced waste disposal costs, improved process efficiency, and enhanced environmental performance – far outweigh the upfront expense.

Careful process design is paramount. Implementing closed-loop systems, where waste streams are recycled and reused within the process, is a powerful strategy. This circular approach minimizes waste sent to disposal and reduces the reliance on virgin materials. Such systems, however, require careful engineering to ensure efficient recycling and reuse without compromising product quality or process safety. Furthermore, optimization of reaction parameters—temperature, pressure, residence time—can minimize byproduct formation and maximize the desired product yield. Precise control over these factors can significantly reduce waste generation.

Technological advancements offer valuable tools for pollution prevention. Membrane separations, for example, can replace energy-intensive distillation or extraction processes, minimizing solvent usage and energy consumption. Microreactor technology, also discussed previously, enables enhanced process control, resulting in higher selectivity and reduced waste. These technological interventions, while requiring initial investment, offer long-term advantages through improved efficiency and reduced environmental impact. Careful evaluation and selection of appropriate technologies are crucial for optimal implementation.

Beyond technological solutions, effective pollution prevention requires a robust commitment to continuous improvement. Regular monitoring and evaluation of process performance, coupled with ongoing assessment of waste generation, are crucial for identifying areas for further improvement. Implementing strategies for continuous process improvement, such as lean manufacturing principles, promotes a culture of waste minimization throughout the organization. This commitment to continuous improvement ensures the long-term success of pollution prevention initiatives and drives sustainable practices into the organization's core. The integration of pollution prevention into routine operations is crucial for its success.

Employee training and engagement are indispensable. A well-informed workforce understands the importance of pollution prevention and is empowered to contribute to its success. Training programs should cover topics like waste identification, safe handling of materials, and implementation of best practices. Engaging employees in the pollution prevention process fosters ownership and commitment, leading to greater success in waste reduction efforts. Investing in employee training is an investment in long-term sustainability.

Regulatory compliance, while often viewed as a burden, offers a valuable framework for pollution prevention. Understanding and adhering to environmental regulations ensures that processes are designed and operated in an environmentally responsible manner. Furthermore, keeping abreast of evolving regulations and best practices provides opportunities to implement new technologies and strategies for improved environmental performance. Compliance is not merely a box to tick; it is a foundation for environmental stewardship.

Finally, pollution prevention is not an isolated endeavor; it's an integral component of a comprehensive sustainability strategy. It aligns seamlessly with the broader principles of green chemistry and life cycle assessment discussed earlier in this book. By reducing waste at its source, pollution prevention significantly minimizes the environmental impact of chemical processes, contributing to a more sustainable future for the chemical industry. The long-term success of pollution prevention depends on its integration into the overall business strategy and commitment to a culture of continuous improvement.

3.5 Clean Production Technologies

Clean production technologies represent a paradigm shift, moving beyond mere waste treatment to proactively minimize environmental impact throughout the entire production lifecycle. This proactive approach integrates environmental considerations into the design and operation of chemical processes, resulting in significantly reduced pollution and resource consumption. It's not just about cleaning up the mess; it's about preventing it from ever happening.

Unlike end-of-pipe solutions that address pollution after it's created, clean production technologies focus on preventing pollution at its source. This requires a fundamental change in mindset, shifting from a linear "take-make-dispose" model to a circular economy where resources are valued and reused. This integrated approach, involving all stages from raw material sourcing to product disposal, emphasizes efficiency and minimizes environmental burdens.

Implementing clean production often involves a combination of strategies tailored to specific processes and industries. For example, process modifications might incorporate better reaction control, leading to higher yields and less waste. Alternative feedstocks, derived from renewable sources, can drastically reduce reliance on fossil fuels and minimize the environmental footprint of the raw materials. Furthermore, efficient separation and purification techniques, such as membrane technology or supercritical fluid extraction, reduce energy consumption and solvent use.

The adoption of clean production technologies can deliver significant economic benefits, beyond environmental advantages. Reduced waste translates to lower disposal costs and potential recovery of valuable byproducts. Improved efficiency often leads to lower energy and raw material consumption, resulting in substantial cost savings. Moreover, a strong commitment to sustainability enhances a company's reputation, strengthens its brand image, and attracts environmentally conscious consumers and investors. However, integrating clean production technologies isn't without challenges. Implementing these changes often requires significant upfront investment in new equipment, processes, and employee training. There can be initial operational hurdles as teams adapt to new methodologies. Additionally, the availability of suitable technologies and expertise can vary across industries and geographical locations, posing a barrier for some companies. Overcoming these challenges requires careful planning, strong management commitment, and collaboration among stakeholders.

A crucial aspect of successful clean production implementation involves comprehensive training programs for personnel. Employees need a thorough understanding of the new technologies and procedures to ensure safe and efficient operation. Regular monitoring and data analysis are essential for tracking progress and identifying areas for improvement. Moreover, regular audits provide valuable insights into environmental performance and compliance, helping companies continuously optimize their processes.

Successfully implementing clean production often necessitates collaboration with external partners, including government agencies, research institutions, and other companies. This collaborative approach leverages external expertise, facilitates knowledge sharing, and encourages the development of new technologies and best practices. Government incentives and policies play a significant role in encouraging the adoption of cleaner production technologies through tax breaks, grants, or regulatory frameworks.

Such support is vital, especially for smaller businesses that may lack the resources to implement changes independently.

The long-term benefits of clean production far outweigh the initial challenges. By minimizing waste, conserving resources, and reducing pollution, companies contribute to environmental sustainability and secure a more secure future for their operations. Clean production technologies are not just an environmental imperative; they're a strategic business advantage that fosters innovation, enhances profitability, and strengthens a company's position in the marketplace. Embracing these technologies is not just about fulfilling regulatory requirements; it's about building a more resilient and sustainable future for business and the planet.

4 Sustainable Solvents & Catalysts

The shift towards sustainable chemical production necessitates a fundamental change in the materials we employ. Traditional solvents, often volatile organic compounds (VOCs), contribute significantly to air pollution and pose health risks. Similarly, many catalysts, while highly effective, are derived from unsustainable sources or exhibit toxicity. Replacing these with greener alternatives is crucial. Ionic liquids, for instance, offer a compelling solution. These salts, liquid at room temperature, possess negligible vapor pressure, eliminating volatile emissions. Their tunable properties, allowing for tailored solvent characteristics through cation and anion selection, further enhance their versatility. Research continues to explore novel ionic liquid structures optimized for specific reactions, improving efficiency and minimizing environmental impact. Their potential extends beyond simple solvent replacement; they can even act as catalysts themselves, streamlining processes.

Bio-based catalysts represent another significant advancement. Derived from renewable sources like plants and microorganisms, these catalysts reduce reliance on fossil fuels and minimize the carbon footprint of chemical production. Enzymes, for example, are highly selective catalysts that operate under mild conditions, reducing energy consumption and byproduct formation. Their inherent biodegradability also simplifies waste management. However, challenges remain in scaling up biocatalyst production and ensuring their stability and activity under industrial conditions. Ongoing research focuses on enzyme engineering and immobilization techniques to overcome these limitations, making bio-based catalysts a practical alternative for a wide range of applications.

Consider the synthesis of fine chemicals, a sector traditionally reliant on hazardous solvents and catalysts. The transition to ionic liquids and bio-based alternatives can significantly reduce the environmental burden of these processes. For example, replacing a VOC solvent with a suitable ionic liquid can eliminate volatile emissions, improving worker safety and reducing air pollution. Similarly, substituting a metal-based catalyst with a biocatalyst can minimize waste generation and reduce the risk of heavy metal contamination. The economic advantages are also noteworthy; reducing waste and improving efficiency can lead to significant cost savings, making sustainable alternatives increasingly attractive to manufacturers.

A comprehensive lifecycle assessment (LCA) is critical when evaluating the sustainability of solvents and catalysts. While a bio-based catalyst may seem inherently greener, its production process and eventual disposal must also be considered. A full LCA evaluates the environmental impact across the entire lifespan of the material, from raw material extraction to waste disposal. This holistic approach ensures that the overall environmental benefit is maximized. It's also essential to consider the energy required to synthesize and purify both ionic liquids and bio-catalysts. Efficient synthesis routes and energy-efficient separation techniques are crucial to minimize the environmental footprint of these greener alternatives. Optimization of reaction conditions can also reduce energy demands during the chemical process itself.

Furthermore, the implementation of sustainable solvents and catalysts is not a standalone effort but part of a broader strategy for green chemistry. Integrating these advancements with other sustainable practices, such as process intensification and waste minimization, creates a synergistic effect, maximizing environmental benefits. The success of this transition relies on collaboration between chemists, engineers, and policymakers. Open communication and knowledge sharing are crucial for accelerating the adoption of sustainable alternatives across various industries. Support from regulatory bodies is equally vital to incentivize the development and implementation of greener technologies. Government incentives and stricter regulations on hazardous materials can significantly influence the market adoption of sustainable solvents and catalysts, encouraging a more environmentally responsible chemical industry. The path towards a truly sustainable chemical industry requires a multifaceted approach that encompasses technological innovation, policy support, and a collective commitment to environmental stewardship.

4.1 Ionic Liquids

Ionic liquids, often abbreviated as ILs, are salts that exist as liquids at or near room temperature. Unlike traditional organic solvents, they possess unique properties making them attractive alternatives in various chemical processes, particularly those striving for environmental sustainability. Their negligible vapor pressure eliminates volatile organic compound (VOC) emissions, a major contributor to air pollution. Furthermore, their tunable properties, achieved by varying the cation and anion components, allow for tailored solvent characteristics ideal for specific reactions and separations. This remarkable versatility is a key advantage, enabling optimization for particular applications.

Their high thermal stability allows for operation at elevated temperatures, potentially enhancing reaction rates and process efficiency. This feature contributes significantly to energy savings, a crucial aspect of sustainable chemical production. Many ILs exhibit excellent solvating capabilities for a wide range of compounds, including both polar and non-polar substances, surpassing the limitations of traditional solvents. This enhanced solvation power can lead to improved reaction yields and selectivity, optimizing resource utilization and reducing waste.

However, the synthesis of certain ILs can involve the use of less sustainable reagents, and the potential toxicity of some ILs must be carefully evaluated. Life cycle assessments are therefore crucial to fully understand their environmental impact, accounting for raw material acquisition, synthesis, application, and eventual disposal or recycling. Careful selection of ILs is vital, prioritizing those with low toxicity and readily available, renewable starting materials.

The potential for recycling and reuse of ILs is a significant advantage in terms of waste minimization. In many cases, ILs can be recovered and purified for subsequent use, dramatically reducing their overall environmental footprint. This reusability significantly contributes to the economic viability of implementing these solvents within industrial settings, reducing both waste and production costs. The development of efficient recycling processes is an active area of research, striving to further enhance the sustainability of IL applications.

Applications of ILs span various chemical processes. They are effective in catalysis, acting as both solvents and catalysts or catalyst supports. Their ability to dissolve both reactants and catalysts promotes enhanced reaction rates and selectivity. In separation processes, ILs are used in liquid-liquid extraction, offering a green alternative to conventional organic solvents in separating valuable products from reaction mixtures or purifying streams. Their ability to selectively dissolve target compounds while leaving impurities behind makes them powerful tools for purification.

The design and selection of appropriate ILs require consideration of several factors. The chemical structure of both the cation and anion influences the IL's properties, including its polarity, viscosity, and solvating ability. Computational tools are increasingly employed to predict these properties, reducing the need for extensive experimental screening. Careful consider-

ation of these properties is vital to ensuring the IL's suitability for the intended application and its compatibility with other process components.
Beyond their direct applications, ILs are incorporated into hybrid materials, enhancing their functionalities. For instance, their combination with other materials creates advanced materials with improved properties like enhanced catalytic activity, selectivity, or stability. These hybrid systems often present synergistic effects, combining the benefits of the IL with those of the other components. This approach opens up new avenues for creating highly efficient and sustainable materials for chemical processing.
The development and implementation of ILs within chemical processes require careful consideration of both technical and economic aspects. The cost of IL synthesis and their recyclability are key factors influencing the overall economic feasibility. Scaling up IL-based processes from laboratory to industrial scales requires addressing challenges like viscosity and mixing efficiency, often requiring specialized equipment or modified process designs. These considerations highlight the importance of integrating economic and technical expertise during IL technology development and implementation.
Ongoing research focuses on developing new ILs with enhanced properties and exploring novel applications. This includes designing ILs with improved biodegradability and reduced toxicity, addressing environmental concerns. The exploration of ILs derived from renewable resources is also a critical area of focus, furthering the integration of green chemistry principles. The continued development of advanced characterization techniques

helps to better understand IL behavior and improve their design.

In conclusion, ionic liquids represent a significant advancement in green chemistry, offering a viable path toward environmentally sustainable chemical production. Their unique properties, including negligible vapor pressure, tunable characteristics, high thermal stability, and excellent solvation capabilities, present significant advantages over traditional solvents. However, careful consideration of factors such as toxicity, recyclability, and economic aspects are crucial for their successful and sustainable implementation in industrial settings. Continuous research and development in this area will further expand their applications and drive the transition toward a greener chemical industry.

4.2 Bio-based Catalysts

Bio-based catalysts offer a compelling pathway towards greener chemical processes, aligning perfectly with the principles of sustainable manufacturing. Their inherent renewability and often reduced toxicity compared to traditional metal-based catalysts represent a significant advancement in environmental stewardship. Imagine a world where industrial processes rely on enzymes derived from readily available biomass, instead of relying on precious metals and their complex, energy-intensive extraction methods. This shift offers not only environmental benefits but also potential economic advantages. The development of bio-based catalysts hinges on understanding their specific properties and carefully tailoring their application to suit various chemical transformations.

The versatility of bio-catalysts is remarkable. Enzymes, for instance, exhibit exquisite selectivity, catalyzing reactions with exceptional precision. This specificity minimizes the formation of unwanted by-products, thereby reducing waste generation and the need for downstream purification steps. This contrasts sharply with many traditional catalysts that often lead to complex reaction mixtures requiring extensive separation and treatment. Furthermore, enzymes often function under milder reaction conditions – lower temperatures and pressures – leading to significant energy savings. These milder conditions not only conserve energy but also increase the safety of chemical processes, reducing the risk of hazardous incidents. Consider the potential for reducing energy consumption and improving safety profiles in industrial settings through this technology.

However, the application of bio-based catalysts isn't without its challenges. Enzyme stability, for example, can be a significant hurdle. Enzymes, being biological molecules, can be sensitive to changes in temperature, pH, and the presence of inhibitors. This necessitates careful process design and optimization to maintain enzyme activity and prevent degradation during the reaction. Furthermore, the production and purification of enzymes can sometimes be complex and expensive, limiting their widespread adoption in certain industrial settings. Researchers are actively addressing these challenges through protein engineering and the development of robust enzyme immobilization techniques. Immobilization, for instance, can increase enzyme stability and allow for easier recovery and reuse, mitigating cost concerns.

One promising avenue of research focuses on modifying existing enzymes or engineering novel ones with enhanced catalytic properties. Directed evolution and rational design are powerful tools for tailoring enzymes to specific applications, improving their activity, selectivity, and stability under industrially relevant conditions. This also allows for the creation of biocatalysts to perform tasks not achievable with naturally occurring enzymes. This research area opens up possibilities for tailoring enzymes to specific chemical reactions and industrial demands. For example, imagine tailoring an enzyme to break down a recalcitrant waste stream for efficient bioremediation.

The economic aspects of transitioning to bio-based catalysts should not be underestimated. While initial investment in research and development might be substantial, the long-term benefits are promising. Reduced waste disposal costs, energy savings, and the potential for creating value-added products from biomass can contribute to a more sustainable and cost-effective chemical production process. Furthermore, a move towards bio-based catalysts aligns with the growing demand for environmentally friendly products and processes, creating market opportunities for businesses embracing this technology. This transition towards sustainable practices can yield economic benefits while promoting environmental protection.

Ultimately, the successful implementation of bio-based catalysts requires a multidisciplinary approach. Chemists, biochemists, engineers, and economists must work together to overcome the challenges and realize the full potential of this transformative technology. Through continuous innovation and collaboration, we can pave the way for a future where chemical manufacturing is both efficient and environmentally responsible. This interdisciplinary collaboration is critical to integrating bio-catalysts effectively into industrial processes. The future of sustainable chemical production relies on such cooperative efforts and the effective merging of scientific knowledge and engineering practices. This collaborative approach is vital in ensuring the successful transition towards a more sustainable chemical industry.

5 Energy Efficiency in Chemical Processes

Reducing energy consumption is paramount in modern chemical manufacturing. We can achieve significant improvements through meticulous process optimization, focusing on reaction conditions, such as temperature, pressure, and reactant ratios. Careful selection of catalysts can dramatically reduce energy requirements by lowering activation energy and speeding reaction rates. For example, the use of heterogeneous catalysts, often more easily separated and reused, minimizes the energy intensive purification steps needed for homogeneous catalysts. Furthermore, advanced process control systems, utilizing real-time data analysis and predictive modeling, allow for fine-tuning operations, minimizing energy waste associated with inefficient control loops. This sophisticated approach optimizes energy use while maintaining product quality.

Heat integration presents a powerful strategy to significantly reduce energy demand in chemical processes. Waste heat, inevitably generated in many chemical reactions, often represents a substantial energy loss. Heat integration techniques recover this valuable energy, transferring it to other parts of the process where heat is needed, such as preheating reactants or supplying energy to distillation columns. Pinch analysis, a sophisticated optimization tool, helps identify the best points of heat exchange, maximizing energy recovery and minimizing utility consumption. Implementing heat exchanger networks efficiently requires careful consideration of temperature profiles, heat capacity flows, and the economic viability of different heat integration options. Effective implementation demands a thorough understanding of the overall process, allowing for a strategic approach to energy recovery.

Renewable energy sources are increasingly becoming essential for sustainable chemical production. Traditional reliance on fossil fuels contributes significantly to greenhouse gas emissions. Transitioning to renewable energy sources, such as solar, wind, and geothermal energy, directly reduces the carbon footprint of chemical manufacturing. Solar thermal technology, for instance, can provide the necessary heat for endothermic reactions. Furthermore, integrating renewable electricity into the process through electrochemical methods can drive energy-intensive reactions. However, the intermittent nature of some renewable energy sources requires careful planning and the possible integration of energy storage solutions, like batteries or thermal storage, to ensure continuous operation. This approach requires a balanced assessment of initial investments versus long-term operational cost savings and environmental benefits. It's not just about substituting energy sources; it's about strategically integrating them for optimal performance and sustainability.

Process integration and optimization are not isolated strategies; they are synergistic elements of a holistic approach to energy efficiency. Consider, for instance, a chemical plant employing heat integration and running on renewable electricity. By combining these strategies, the plant not only reduces its carbon footprint but also optimizes its energy usage, leading to substantial cost savings. These cost savings can then be reinvested into further sustainability initiatives or into other research and development efforts. This integrated perspective transcends individual optimization efforts to create a robust system that's both environmentally responsible and economically viable. The success of this holistic approach relies heavily on detailed process modeling and simulation, followed by careful implementation and monitoring to fine-tune operations for optimal performance.

Beyond the technical aspects, successfully integrating energy efficiency measures requires careful consideration of the economic factors and broader societal implications. The initial investment in new technologies, such as advanced process control systems or renewable energy installations, can be significant. Therefore, a thorough life cycle cost analysis, comparing initial capital costs with ongoing operational savings and environmental benefits, is crucial. This necessitates a balanced approach, taking into account energy costs, reduced waste disposal fees, potential carbon tax credits, and the overall positive impact on the company's image. Moreover, engaging with the local community and complying with relevant environmental regulations are equally important aspects of achieving sustainable and responsible chemical production. The full picture includes more than technical advancements; it's a comprehensive strategy focused on environmental responsibility, economic viability, and social awareness.

5.1 Process Optimization

Process optimization is crucial for environmentally sound chemical manufacturing. It's about identifying bottlenecks, inefficiencies, and areas ripe for improvement within existing processes, leading to substantial reductions in energy consumption, waste generation, and overall environmental impact. We aren't talking about minor tweaks; we're discussing a systematic approach to redesigning processes for maximum sustainability.

Consider a classic example: a batch reactor process. Initial analysis might reveal long reaction times, excessive energy input for heating and cooling cycles, and significant solvent waste. Through process optimization, we could explore alternative reactor designs, like continuous flow reactors, that offer better temperature control, faster reaction kinetics, and reduced solvent volumes. Careful modeling and simulation, employing software capable of predicting process behavior under different conditions, become essential tools.

Implementing advanced process control systems plays a vital role. These systems, utilizing real-time data and sophisticated algorithms, allow for precise adjustments to operating parameters, ensuring optimal performance while minimizing deviations that could lead to increased waste or energy consumption. Think of it as a self-regulating system constantly striving for efficiency. Such precision reduces the likelihood of off-spec products needing reprocessing, another contributor to environmental burden.

Beyond reactor design and control, optimization extends to other stages of the chemical process. Separation processes, for example, are often energy-intensive. Analyzing existing techniques—distillation, extraction, filtration—with a critical eye often unveils opportunities for improvement. Implementing more efficient separation methods, such as membrane technologies or supercritical fluid extraction (covered in a previous chapter), can significantly reduce energy demands and waste.

The optimization process itself benefits from a structured approach. Begin with a thorough process mapping exercise, identifying all the steps involved in the chemical transformation, from raw material acquisition to final product packaging. This visualization helps to pin point points of inefficiency. Then, utilize data analysis techniques, drawing on historical production data, to identify trends and patterns that reveal areas for potential improvement. This data-driven approach is indispensable for making informed decisions.

Don't underestimate the power of teamwork in process optimization. It requires collaboration between chemical engineers, environmental scientists, and process operators. Their combined expertise ensures a holistic perspective and aids in the identification of potentially overlooked opportunities. Open communication and regular performance reviews are crucial for maintaining the momentum of this continuous improvement process.

Furthermore, exploring alternative reaction pathways is an often-overlooked aspect of optimization. A seemingly minor adjustment to the reaction mechanism, perhaps by altering the catalyst or introducing a co-solvent, can significantly affect reaction rates, selectivity, and overall efficiency. This step usually involves a deeper understanding of reaction kinetics and thermodynamics. Remember, small changes can have a big impact on sustainability.

Beyond the technical aspects, consider the economic implications. While the initial investment in optimization strategies might seem substantial, the long-term benefits often far outweigh the costs. Reductions in energy consumption, waste disposal fees, and raw material usage directly translate into financial savings. And increasingly, environmentally conscious consumers and stricter regulations favor companies demonstrating a commitment to sustainability. This translates into improved brand image and access to new markets.

Finally, continuous monitoring and evaluation are essential for long-term success. Regularly assess the effectiveness of implemented optimizations, tracking key performance indicators (KPIs) such as energy consumption, waste generation, and yield. Use this data to fine-tune strategies over time, adapting to changing market demands and technological advancements. This cyclical approach ensures that your processes remain optimized for maximum environmental stewardship and economic viability. The quest for sustainable chemical manufacturing is a continuous journey, not a destination. Process optimization provides a powerful toolkit to navigate this path effectively.

5.2 Heat Integration

Heat integration is a cornerstone of energy efficiency in chemical processing, offering substantial economic and environmental benefits. It involves strategically connecting process streams to recover and reuse waste heat, minimizing the need for external heating or cooling. This clever approach reduces energy consumption, lowers operational costs, and significantly diminishes a plant's carbon footprint—a critical aspect of environmentally responsible manufacturing. Implementing heat integration requires a systematic approach, starting with a thorough understanding of the process's thermal profile.

Identifying potential heat sources and sinks is crucial. A heat source, such as an exothermic reactor effluent, possesses excess heat that can be harnessed. Conversely, a heat sink, like an endothermic reactor requiring heating, represents a potential recipient for this recovered energy. Careful analysis of temperature profiles, flow rates, and heat capacities of these streams is paramount in determining the feasibility of heat integration. This detailed analysis often involves the use of specialized software tools that simulate process behavior and optimize heat exchanger network design.

Once potential matches between heat sources and sinks are identified, the design of the heat exchanger network (HEN) begins. This network, typically composed of several heat exchangers, facilitates the transfer of heat between streams. The optimization of the HEN is a complex task, requiring consideration of factors such as temperature differences, pressure drops, and the economics of different heat exchanger types. Efficient HEN design minimizes the number of exchangers, reducing capital costs and operational complexity. Various optimization algorithms and software packages are available to aid in this process, improving efficiency and minimizing material costs.

Beyond the technical aspects of designing and implementing an HEN, the practical implications are equally significant. Successful integration necessitates close collaboration between process engineers, instrumentation specialists, and operations personnel. This ensures proper installation, commissioning, and ongoing maintenance of the heat exchange equipment. Regular monitoring of the system's performance, through instrumentation and data analysis, is vital for identifying potential issues and maintaining optimal efficiency. Training operational personnel on the nuances of the new system is also crucial for ensuring safe and effective operation.

The economic benefits of heat integration extend beyond reduced energy costs. Decreased energy consumption translates into lower emissions, potentially enhancing a company's environmental, social, and governance (ESG) profile. This positive environmental impact can improve a company's brand reputation and attract environmentally conscious investors. The potential reduction in greenhouse gas emissions can help meet regulatory compliance requirements and potentially gain access to carbon offsetting schemes or government incentives. Moreover, efficient energy utilization improves the overall profitability of the chemical production processes, leading to higher margins.

Finally, integrating heat recovery into existing plants may pose operational challenges. Retrofitting a heat integration system into an already functioning facility often involves significant modifications and careful planning. The system's compatibility with existing infrastructure must be thoroughly assessed and managed. A thorough risk assessment is also essential to anticipate potential safety concerns associated with handling hot fluids and the installation of new equipment. Successful implementation requires careful scheduling, potentially involving temporary production shutdowns, and coordination with plant maintenance teams. Detailed planning is paramount to minimise disruptions and ensure a smooth transition. Ignoring these aspects can lead to higher costs and potential delays, highlighting the importance of a well-structured and thoroughly planned implementation process. The long-term rewards—enhanced environmental performance and significant cost savings—clearly outweigh these initial hurdles.

5.3 Renewable Energy Sources

Harnessing renewable energy sources presents a crucial opportunity to drastically reduce the environmental impact of chemical processes. Solar energy, with its abundant and readily available nature, offers a compelling alternative to fossil fuels. Photovoltaic systems directly convert sunlight into electricity, powering various aspects of chemical manufacturing, from reactor operation to equipment maintenance. Concentrated solar power (CSP) systems, using mirrors to focus sunlight, generate high-temperature heat suitable for endothermic reactions, further improving energy efficiency and reducing reliance on traditional heating methods. The intermittent nature of solar power can be mitigated through energy storage solutions, such as batteries or pumped hydro, ensuring a continuous supply of clean energy.

Wind energy, another significant renewable resource, provides a consistent energy stream, especially in regions with high wind speeds. Wind turbines convert kinetic energy into electricity, which can be seamlessly integrated into the chemical plant's power grid. Offshore wind farms, offering stronger and more consistent winds, have particularly high potential for large-scale chemical production facilities located near coastlines. Careful site selection is critical to maximize energy output while minimizing environmental disturbances associated with wind turbine placement. Like solar, energy storage is key to addressing the intermittency of wind power.

Hydropower, leveraging the energy of flowing water, is a mature renewable technology offering substantial potential for integrating sustainability into chemical manufacturing. Hydroelectric dams generate electricity by harnessing the potential energy of water stored behind the dam. Run-of-river hydropower plants, with minimal environmental impact, are particularly attractive for smaller-scale chemical operations. These plants harness the kinetic energy of flowing water without creating large reservoirs, leading to lower ecological disruption compared to traditional dam systems. The reliability of hydropower, combined with its established infrastructure in many regions, makes it a highly attractive option.

Bioenergy, derived from biomass, is another crucial player in reducing our chemical industry's carbon footprint. Biomass, encompassing organic materials like agricultural residues and forestry products, can be converted into biofuels through various processes, such as gasification or pyrolysis. These biofuels can then replace fossil fuels in heating and power generation within chemical plants. Furthermore, utilizing waste biomass from the agricultural and forestry sectors fosters a circular economy, minimizing waste while simultaneously providing a sustainable energy source. Careful consideration must be given to the sustainability of biomass sourcing to avoid deforestation or other negative environmental consequences.

Geothermal energy, tapping into the Earth's internal heat, offers a reliable and consistent baseload power source for chemical processes. Geothermal power plants utilize steam or hot water extracted from underground reservoirs to generate electricity. This technology is particularly valuable in regions with high geothermal activity, providing a consistent energy supply irrespective of weather conditions. The relatively low environmental impact of geothermal energy, coupled with its reliability, is a major advantage for sustainable chemical production. However, geothermal resources are geographically limited, restricting its widespread applicability. The careful management and monitoring of geothermal wells are vital to prevent ground subsidence or other environmental issues.

Integrating these renewable energy sources requires a multifaceted approach. Careful planning is essential to assess the availability and suitability of each resource for a specific chemical facility's location and energy needs. Energy storage solutions are vital to manage the intermittency of solar and wind power. Smart grids, enabling real-time monitoring and control of energy distribution, are crucial for efficient integration and optimization of renewable energy sources within chemical production plants. The transition toward renewable energy may necessitate initial investments in new infrastructure, but the long-term environmental and economic benefits far outweigh the upfront costs, paving a path toward truly sustainable chemical manufacturing. Furthermore, government policies and incentives can play a key role in accelerating this crucial transition to a cleaner and more sustainable chemical industry, encouraging investment in renewable energy technologies and incentivizing their adoption by chemical companies. This collaborative approach—integrating technological advancements, economic incentives, and supportive policy frameworks—will be key to decarbonizing the chemical industry.

6 Life Cycle Assessment (LCA)

Life cycle assessment, or LCA, is a powerful tool for evaluating the environmental impact of a product or process across its entire lifespan, from raw material extraction to disposal. It's not simply about measuring emissions at the end of a production line; instead, LCA considers the cumulative effects of all stages, providing a comprehensive picture of environmental burdens. This holistic approach is crucial for informed decision-making in chemical process refinement.

Understanding the LCA methodology is paramount. The process typically involves four key stages: goal and scope definition, inventory analysis, impact assessment, and interpretation. Defining the study's purpose and boundaries is the first crucial step. This includes specifying the functional unit (e.g., the production of one kilogram of a specific chemical), the system boundaries (what is included and excluded from the analysis), and the allocation methods if multiple products are generated from a single process. Failing to define these parameters clearly can lead to inaccurate or misleading results.

Following goal definition, inventory analysis involves meticulously quantifying all energy and material inputs and outputs across all stages of the product's lifecycle. This detailed inventory includes raw materials, energy consumption, emissions to air, water, and soil, and waste generation. Data collection can be challenging, requiring access to reliable process data, material property information, and emission factors. Often, this stage relies heavily on life cycle databases and modelling software to accurately account for all inputs and outputs. Data quality is paramount, and uncertainties inherent in the data should be carefully addressed.

Once the inventory is compiled, an impact assessment is conducted. This stage translates the inventory data into meaningful environmental indicators, such as global warming potential, ozone depletion potential, and acidification potential. Various impact assessment methods exist, each with its strengths and weaknesses. The choice of method depends on the study's objectives and the available data. A thorough impact assessment highlights the most significant environmental hotspots in the product or process lifecycle, allowing for targeted improvement efforts.

The final step, interpretation, involves analyzing the results of the impact assessment to identify key environmental issues and inform decision-making. This stage requires careful consideration of data uncertainties and limitations, as well as potential biases associated with the chosen methods and data sources. Clear and concise communication of LCA findings is crucial, ensuring that stakeholders understand the implications of the assessment for environmental performance improvement. Effective communication may involve the use of visual aids, such as charts and graphs, and the tailoring of the report to the specific audience. This allows for actionable feedback that improves future iterations of the chemical process.

Effective LCA relies heavily on accurate data, and this can pose a significant challenge. Obtaining comprehensive data across all stages of a product's lifecycle can be difficult and time-consuming. Furthermore, the choice of data sources and assumptions can influence the outcome of the analysis. Therefore, transparent documentation of data sources and methodologies is essential to ensure the credibility and reliability of the LCA results.

Data limitations can also lead to uncertainty in the results, and it is crucial to quantify and address these uncertainties during the interpretation phase. The use of sensitivity analysis helps to understand how variations in input data might affect the overall results. This approach enhances the robustness and reliability of LCA findings, strengthening their impact on informed environmental decision-making.

Beyond the technical aspects of conducting an LCA, effective communication is critical for maximizing its impact. Presenting the findings clearly and concisely to both technical and non-technical audiences is essential. This involves using visual aids and tailoring the presentation to match the knowledge level and interests of the audience. The aim is to translate complex environmental data into actionable insights that drive improvements in the design and management of chemical processes.

By thoroughly understanding the LCA methodology and its limitations, chemical engineers and other professionals can use it as a crucial tool for designing and evaluating sustainable chemical processes. This involves not only minimizing environmental impacts but also optimizing resource efficiency, promoting circular economy principles, and fostering responsible environmental stewardship across the entire life cycle. The integration of LCA within the chemical process development process ensures that environmental considerations are addressed throughout the lifecycle of a product or process, leading to more sustainable outcomes and a reduced environmental footprint. This proactive approach is essential for building a future where chemical production thrives without compromising environmental integrity.

6.1 LCA Methodology

Conducting a Life Cycle Assessment (LCA) involves a systematic approach to evaluating the environmental impacts associated with a product or process throughout its entire life cycle. This encompasses everything from raw material extraction and processing to manufacturing, transportation, use, and ultimately, disposal or recycling. The process is meticulously structured, demanding careful consideration at every stage.

A well-defined goal and scope are paramount. Before initiating the assessment, you need to clearly articulate the product or process under scrutiny, the specific environmental impacts of interest (e.g., greenhouse gas emissions, resource depletion, toxicity), and the geographical boundaries of the analysis. This initial stage sets the foundation for a rigorous and meaningful study, guiding subsequent steps. Failure to establish a clear scope often leads to inconsistencies and difficulties in interpretation.

The inventory analysis forms the core of the LCA. Here, all inputs and outputs associated with each life cycle stage are meticulously quantified. This involves data collection on energy consumption, raw material usage, emissions to air, water, and soil, and waste generation. Data sources can range from company records and material safety data sheets to published literature and databases, demanding careful scrutiny and validation. The accuracy of the inventory directly influences the reliability of the subsequent impact assessment.

Impact assessment involves translating the inventory data into meaningful environmental indicators. This step necessitates using established methodologies and impact assessment models, such as those provided by the IMPACT World+ or ReCiPe frameworks. These models quantify the environmental burdens associated with various pollutants and resource consumption, enabling a comparative analysis of different products or processes. Choosing the appropriate model hinges on the specific environmental impacts prioritized in the study's scope.

Interpreting the LCA results requires careful consideration. The results should not be viewed in isolation; you must analyze them within the defined scope and limitations of the study. Consider the uncertainties associated with the data, acknowledging that some data points may be less reliable than others. Effective communication of the findings is key, often involving visually appealing charts and graphs that convey complex data in an accessible manner. This facilitates informed decision-making and fosters a clearer understanding of the environmental implications.

Beyond the core stages, several key considerations enhance the robustness of an LCA. Sensitivity analysis helps assess the influence of uncertainties in the data on the overall results. This process involves systematically varying input parameters to gauge their effect on the impact scores. By identifying the most influential parameters, you can focus future data collection efforts, improving the accuracy of the assessment.

Furthermore, the inclusion of life cycle costing (LCC) offers a valuable complement to the environmental assessment. LCC analyses the economic aspects of the various life cycle stages, providing a holistic perspective on the financial implications of sustainability choices. Integrating LCC with LCA provides decision-makers with a comprehensive understanding of both environmental and economic performance.

The selection of appropriate functional units is crucial. A functional unit defines the reference unit of the product or process being assessed (e.g., kilogram of product, kilometer traveled). Consistency in functional units across different LCAs is critical for comparative analysis. Inconsistent choices can lead to skewed interpretations, undermining the value of the study. Precise definition ensures the validity of your comparisons.

In conclusion, a well-executed LCA provides a powerful tool for evaluating the environmental performance of chemical processes. While demanding in terms of data collection and analysis, the insights it yields are invaluable for promoting sustainable practices in the chemical industry. By meticulously following the methodology outlined here, practitioners can make informed decisions, contribute to environmental stewardship, and foster innovation towards greener chemical production. Remember, the accuracy and validity of your LCA hinge on the thoroughness of each step; meticulous attention to detail translates into reliable results that guide responsible actions.

6.2 Interpreting LCA Results

Understanding an LCA's output requires more than just numbers; it demands a nuanced interpretation. Raw data, like energy consumption or greenhouse gas emissions, are meaningless without context. Consider the chosen functional unit – is it representative of the product's actual use? An inaccurate functional unit can skew results significantly, painting a misleading picture of environmental impact. For instance, comparing a plastic bag's impact based on weight versus its number of uses will lead to dramatically different conclusions. The scope also matters—a cradle-to-gate analysis differs drastically from cradle-to-grave, the former omitting crucial downstream impacts. A comprehensive analysis is essential, considering all stages, from raw material extraction to disposal, to accurately reflect the complete environmental burden. This necessitates collaboration across departments and potentially even with external stakeholders, to gain access to all necessary data. Failing to account for all relevant factors could result in an inaccurate reflection of the overall environmental profile.

The choice of impact assessment methodology significantly influences the results. Different methods prioritize various environmental concerns, resulting in potentially conflicting conclusions. For example, a method focused solely on greenhouse gases may overlook other important impacts, like water pollution or resource depletion. Therefore, a critical review of the chosen methodology and its inherent limitations is crucial. Sensitivity analysis helps identify the most influential parameters affecting the outcome. This analysis allows for a clearer understanding of the uncertainties associated with the data and helps quantify the robustness of the conclusions reached. A well-executed sensitivity analysis strengthens the overall reliability and credibility of the results. Choosing appropriate indicators and reporting methods is another crucial step. Consider the target audience – a technical report requires different detail than a summary for management. Visual representations, such as graphs and charts, can significantly improve understanding. They can highlight key contributions to the overall environmental impact and show the effectiveness of different mitigation strategies. However, clear labeling and explanations are paramount to avoid misinterpretations. Effective communication of results is vital to influence decision-making. Presenting the findings in a clear and concise manner, tailored to the audience, is critical for effective communication of the research and its outcomes. Failure to do so can lead to the information being misconstrued or not acted upon. Think about engaging stakeholders in interactive workshops. This allows for deeper engagement and a better understanding of the results and their implications. Always remember your audience. Your findings must be understood by them if you want to drive change. This includes understanding potential limitations and uncertainties of the data involved.

Beyond the numerical data, the interpretation stage of LCA involves critical thinking and professional judgment. Identifying significant environmental hotspots, stages of the process where the most significant impact occurs, is vital for prioritizing areas for improvement. Are the results consistent with expectations? Do the findings align with previously published research in similar areas? This comparative analysis is essential for gaining confidence in the results' validity and making them more meaningful. A gap analysis comparing your process against industry best practices or regulatory standards provides valuable insight. This identifies areas for improvement and guides strategic decision-making. Remember that LCA is a tool, not an end in itself. The ultimate goal is to use the insights from the analysis to drive improvements in environmental performance. These insights can guide adjustments to process parameters, changes in raw materials, or the implementation of waste management strategies. Interpreting LCA results is thus an iterative process, constantly refining the analysis to inform sustainability goals. By integrating it into decision-making, LCA strengthens environmental stewardship.

7 Case Studies & Future Directions

The pharmaceutical industry offers compelling case studies. Companies like Pfizer have invested heavily in process analytical technology (PAT) to optimize reaction conditions in real-time, minimizing waste and improving product yield. This real-time monitoring allows for immediate adjustments, preventing off-specification batches and reducing the need for costly rework or disposal. Furthermore, the adoption of continuous manufacturing, a departure from traditional batch processing, offers significant potential for energy savings and reduced environmental impact. The transition involves integrating multiple process steps into a single, continuous flow, increasing efficiency and reducing downtime. However, challenges remain; transitioning existing facilities and training personnel require substantial investment and time commitment. Scaling up continuous processes while maintaining consistent quality and safety can also prove challenging.

Consider the textile industry, a sector notorious for its high water and energy consumption, along with significant wastewater discharge. Innovative approaches are emerging, such as the use of supercritical CO2 as a solvent for dyeing processes. This method drastically reduces water usage, eliminates the need for harsh chemicals, and minimizes wastewater treatment costs. However, the high pressure and specialized equipment needed represent significant upfront investment hurdles. Moreover, the scalability of this technology for various fabric types and dye compositions requires further investigation. Companies are also exploring bio-based dyes derived from natural sources, a move that reduces reliance on petroleum-based chemicals and promotes a more circular economy. The inherent variability in natural dyes, however, requires careful quality control and standardization procedures.

Looking ahead, several emerging technologies hold significant promise for refining chemical processes. Artificial photosynthesis, for instance, aims to mimic natural photosynthesis to produce fuels and chemicals from sunlight and water, reducing reliance on fossil fuels. While still in early stages of development, this technology offers a pathway towards truly sustainable chemical production, minimizing carbon emissions and promoting energy independence. Another area of focus is the development of advanced materials with tailored properties for catalytic applications. These materials can enhance reaction selectivity and efficiency, minimizing waste and energy consumption. For example, research into metal-organic frameworks (MOFs) has shown significant potential in various catalytic processes, including CO2 capture and conversion. However, the cost-effective synthesis and long-term stability of these materials remain active research areas.

Policy and regulation play a vital role in driving the adoption of sustainable chemical processes. Governments worldwide are increasingly implementing stricter environmental regulations, imposing penalties for exceeding emission limits and promoting the development of cleaner technologies through incentives and grants. Carbon pricing mechanisms, such as carbon taxes or emissions trading schemes, incentivize businesses to reduce their environmental footprint. However, the effectiveness of these policies depends on their design and implementation, ensuring a level playing field for businesses and avoiding unintended consequences. Harmonizing regulations across different countries is also crucial to facilitate international collaboration and prevent regulatory arbitrage.

Significant research opportunities exist to further refine and optimize sustainable chemical processes. Research efforts are focused on developing more efficient catalysts, finding alternative feedstocks from renewable resources, and designing processes that minimize waste generation at the molecular level. Advances in process modeling and simulation are facilitating the design and optimization of green chemical processes, reducing the need for extensive laboratory experiments. Moreover, research is essential to improve our understanding of the long-term environmental impacts of new technologies and materials, ensuring their overall sustainability. Life cycle assessments (LCA) are becoming increasingly sophisticated, providing detailed insights into the environmental impacts of chemical processes throughout their entire lifecycle.

Despite the significant progress made, challenges remain in the widespread adoption of sustainable chemical processes. The high upfront costs associated with implementing new technologies can be a barrier for smaller businesses. Lack of awareness and expertise among industry professionals can hinder the effective implementation of sustainable practices. Furthermore, balancing economic viability with environmental protection often presents a complex trade-off, requiring careful consideration of costs, benefits, and potential risks. Overcoming these challenges necessitates collaborative efforts between academia, industry, and government to develop and deploy sustainable technologies, fostering a culture of innovation and environmental stewardship across the chemical industry. The future of chemical production hinges on embracing these challenges and actively pursuing solutions that enhance both economic prosperity and environmental sustainability.

7.1 Industrial Examples

The pharmaceutical industry, a significant contributor to environmental impact, is actively seeking greener alternatives. Consider the synthesis of active pharmaceutical ingredients (APIs): traditional methods often involve hazardous solvents and generate substantial waste. Innovative approaches like continuous flow chemistry, using microreactors and intensified processes, are now being adopted. These systems enable precise control over reaction parameters, minimizing byproduct formation and enhancing yields. This not only reduces waste disposal challenges but also improves overall energy efficiency, offering both economic and environmental benefits. The transition towards continuous processing requires careful optimization, and integration with in-line analytics can further enhance process control and product quality. Companies are actively investing in this area, understanding that environmental responsibility is increasingly critical for market access and brand reputation. Furthermore, the increased efficiency often leads to a more competitive cost structure in the long run.

A contrasting example lies in the textile industry, a major polluter due to water consumption, chemical usage, and wastewater discharge. Sustainable practices are transforming the industry. For instance, the adoption of supercritical CO2 extraction for dyeing processes offers a solvent that is environmentally benign and effective. This eliminates the need for water-intensive processes and reduces the risk of harmful chemicals entering the environment. While the initial investment might be significant, the long-term cost savings and environmental benefits outweigh the immediate expense. Furthermore, consumers are increasingly demanding environmentally responsible products; this shift in consumer preference is driving the transition toward greener manufacturing techniques within the textile sector. Companies that fail to adapt risk losing market share to competitors committed to sustainable practices.

The agricultural sector, vital to global food security, also presents opportunities for green chemistry applications. The synthesis and application of pesticides and fertilizers are often associated with significant environmental impacts, including soil and water contamination. The development of bio-based pesticides derived from natural sources, along with precise application techniques to minimize pesticide drift, is crucial for mitigating these effects. Similarly, improved fertilizer formulations minimize nitrogen runoff, a major source of water pollution. While the development of effective and cost-competitive bio-based alternatives remains a challenge, this field is witnessing growing investment and innovation. The potential long-term benefits, including reduced environmental impact and enhanced food security, make this research direction increasingly attractive for both governmental agencies and private companies. Moreover, the certification and labeling of sustainable agricultural products allow consumers to make informed choices and support environmentally responsible practices.

Let's explore the challenges in the chemical industry itself. The production of bulk chemicals, like plastics and polymers, often relies on energy-intensive processes and generates significant waste streams. The exploration of catalytic processes, employing highly selective and reusable catalysts, can significantly minimize waste generation. The development and implementation of innovative catalytic materials allow for a reduction in energy consumption and enhance the efficiency of chemical reactions. While the initial catalyst development might require considerable research, the downstream cost and environmental savings can be substantial. Moreover, the development and implementation of innovative recycling technologies is gaining momentum, with many companies actively investing in the creation of chemical recycling processes to transform waste plastics into valuable materials. This approach reduces landfill waste and contributes to a circular economy, addressing concerns about the accumulating plastic waste problem.

Finally, the cement industry, a large emitter of greenhouse gases, is undergoing significant changes. Researchers are exploring alternative cementitious materials with lower carbon footprints, such as geopolymers, offering a sustainable alternative to traditional Portland cement. These novel materials demonstrate comparable strength and durability, while significantly reducing CO2 emissions. While further research is needed to optimize their performance and cost-effectiveness, they represent a promising pathway toward decarbonizing this energy-intensive industry. Furthermore, innovative manufacturing techniques and process optimization can improve energy efficiency in cement production and minimize overall emissions. The implementation of carbon capture and storage technologies is also gaining traction, offering another pathway towards reducing greenhouse gas emissions from the cement industry, which plays a substantial role in global carbon emissions.

7.2 Emerging Technologies

Artificial photosynthesis, mimicking nature's process of converting sunlight into chemical energy, holds immense promise. While still in its early stages for industrial-scale applications, advancements in nanomaterials and catalyst design are accelerating its potential. Imagine chemical plants powered by sunlight, drastically reducing reliance on fossil fuels and lowering greenhouse gas emissions. This technology isn't just about powering the plant;

it's about generating the very building blocks of many chemicals directly from sunlight and CO2, closing the carbon loop and creating a truly circular economy. The challenge lies in improving efficiency and scalability to make it economically viable.

Current research focuses on developing more efficient photocatalysts, capable of absorbing a wider range of sunlight wavelengths and exhibiting higher conversion rates. Scientists are exploring various nanomaterials, including titanium dioxide and perovskites, to enhance light absorption and charge separation. Advanced techniques such as surface modification and doping are also being investigated to optimize catalytic activity and stability. The integration of artificial photosynthesis with other green technologies, such as electrochemical processes and bio-based materials, could lead to even more sustainable chemical production methods. This synergy creates opportunities for innovative and holistic solutions.

Furthermore, the development of robust and cost-effective reactor designs is crucial for scaling up artificial photosynthesis. The design of these reactors needs to consider factors such as light penetration, gas-liquid mass transfer, and heat management. Researchers are exploring different reactor configurations, including photobioreactors that combine biological and photochemical processes, offering unique advantages. These optimized reactor designs will not only enhance efficiency but also increase the overall feasibility of integrating artificial photosynthesis into existing industrial processes. The long-term sustainability of this technology relies heavily on the successful development and implementation of scalable and cost-effective reactor designs.

Another burgeoning area is bio-electrochemical systems (BES). BES use microorganisms to facilitate chemical transformations, often coupled with electrochemical processes to enhance efficiency and selectivity. This unique approach harnesses the power of nature's own catalysts—enzymes—to drive chemical reactions. Many researchers are exploring the use of BES for the production of valuable chemicals from renewable biomass resources, reducing reliance on fossil fuel-derived feedstocks. This technique shows a strong potential for creating sustainable chemical pathways, even for complex chemical productions and conversions previously reliant on traditional methods. The careful consideration of microbial communities and careful environmental conditions is critical for successful operation.

Beyond these core technologies, advanced process control and optimization techniques are essential. Real-time monitoring and data-driven decision-making will be critical for maximizing efficiency and minimizing waste in these complex systems. The integration of artificial intelligence and machine learning algorithms can assist in optimizing reactor operation, predicting process performance, and enhancing overall sustainability. By utilizing this advanced technology for data processing and predictive modeling, we gain a tool to understand the chemical processes and make informed decisions in real time. The combination of advanced process control and emerging technologies like AI will be indispensable for pushing these technologies into the mainstream.

Moreover, economic and social factors significantly influence the adoption of any new technology. The cost-effectiveness of these emerging technologies needs to be carefully evaluated, considering both capital and operational expenditures. Life cycle assessments (LCA) can help in assessing the overall environmental footprint and economic viability of each method, allowing for a comprehensive evaluation of sustainability aspects. Public awareness and policy support are also vital in driving the widespread adoption of sustainable technologies. The successful transition to a more sustainable chemical industry demands a concerted effort from researchers, policymakers, and industry stakeholders. It necessitates comprehensive assessments which consider both environmental and financial outcomes.

In conclusion, the future of sustainable chemical production lies in the convergence of these emerging technologies. Artificial photosynthesis, bio-electrochemical systems, advanced process control, and a proactive approach to economic evaluation are all integral parts of the transition. Addressing the challenges related to efficiency, scalability, and economic feasibility is critical. However, the potential environmental and economic benefits make continued investment and research in these areas essential for building a more sustainable future for the chemical industry. The next decade will be pivotal in determining how successful we are at integrating these exciting, potentially transformative, technologies into our processes and practices. This integration will necessitate cross-disciplinary collaboration and innovation to make the dream of a sustainable chemical industry a reality.

7.3 Policy & Regulation

Navigating the complex web of environmental regulations is crucial for sustainable chemical manufacturing. Understanding and complying with these policies isn't just about avoiding penalties; it's about proactively shaping a future where responsible chemical production is the norm, not the exception. This necessitates a deep dive into the specific regulations relevant to your operations, encompassing everything from waste disposal protocols to emissions standards. For instance, the European Union's REACH regulation (Registration, Evaluation, Authorization, and Restriction of Chemicals) demands comprehensive registration of substances and places a significant emphasis on demonstrating the safe use of chemicals. Compliance demands careful documentation, rigorous testing, and potentially, significant investment in process modifications. Ignoring such regulations isn't an option; it invites legal repercussions, reputational damage, and potentially, environmental catastrophes.

Successful navigation hinges on proactive engagement. Don't wait for legislation to catch up; actively anticipate evolving standards. This involves staying informed about legislative changes, industry best practices, and emerging environmental concerns. Regularly reviewing and updating your company's environmental compliance procedures is essential, as regulations constantly evolve. Consider subscribing to relevant industry publications, participating in conferences and workshops, and consulting with environmental law specialists. These proactive measures not only mitigate risk but also establish your company as a responsible environmental steward, boosting your reputation amongst customers and investors.

Furthermore, embracing a culture of continuous improvement is paramount. This means fostering a mindset where environmental compliance is integrated into all aspects of your operations, from research and development to manufacturing and waste management. Regular internal audits can pinpoint areas needing improvement, helping identify potential compliance gaps before they become major issues. Employee training programs focusing on environmental responsibilities are crucial for ensuring everyone understands their role in environmental stewardship. A commitment to continuous improvement demonstrates not only your dedication to environmental protection but also a proactive approach to risk management.

Policy and regulation aren't merely a checklist of requirements; they are the foundation upon which sustainable chemical practices are built. They provide a framework within which innovation can flourish. Many regulations incentivize the adoption of cleaner technologies, offering grants, tax breaks, or other financial incentives for businesses that invest in sustainable practices. Actively seeking and utilizing these incentives can significantly reduce the costs associated with implementing greener technologies, making the transition more economically feasible. This proactive approach transforms regulatory compliance from a burden into an opportunity for both environmental and economic benefits.

Beyond direct compliance, consider actively participating in the policy-making process. Engage with industry associations, government agencies, and other stakeholders to influence the development of future regulations. By contributing to the conversation, you can help shape policies that are both effective and conducive to innovation in sustainable chemistry. This involvement strengthens your voice within the industry and demonstrates your commitment to responsible practices, securing a more sustainable future for your business and the environment alike. This active participation translates into a more robust and effective regulatory landscape, ultimately benefitting the entire chemical industry. Remember, sustainable chemical production isn't simply about complying with rules; it's about embracing a new paradigm of responsible innovation and environmental leadership. Proactive engagement with policy and regulation is a pivotal aspect of this transformation.

7.4 Research Opportunities

The pursuit of truly sustainable chemical processes demands continuous innovation, and that innovation hinges on robust research. Several key areas currently present significant opportunities for impactful research. Developing novel, highly selective catalysts derived from abundant biomass resources represents a particularly promising avenue. This could significantly reduce reliance on precious metals and minimize environmental impact associated with their extraction and processing. Furthermore, research into catalyst deactivation mechanisms and regeneration strategies is crucial for enhancing the long-term viability of these sustainable alternatives. A comprehensive understanding of these mechanisms would enable the development of more resilient catalysts, reducing replacement frequency and associated waste.

Focusing research efforts on the development of advanced process analytical techniques (PAT) offers another substantial opportunity. Real-time monitoring and control capabilities, enabled by such technologies, would allow for precise optimization of reaction parameters, leading to higher yields, reduced waste generation, and improved energy efficiency. Furthermore, the integration of PAT into closed-loop control systems can drastically improve process consistency and reproducibility. This precise control minimizes the need for extensive cleanup or rework, further contributing to sustainability goals. Research should focus on developing PAT methodologies tailored specifically to the needs of green chemical processes, addressing challenges such as the complexity of reaction mixtures and the need for non-invasive, real-time measurements.

Another significant research area lies in the development and application of life cycle assessment (LCA) methodologies specifically designed for evaluating the sustainability of chemical processes. Current LCA practices often lack the detailed chemical-specific data needed to accurately assess the impacts of specific chemical transformations. More sophisticated modeling techniques, coupled with advanced experimental data, are necessary for more precise LCA studies. This is essential for accurately identifying the most impactful stages of a process and for prioritizing the most effective strategies for improvement. Researchers can improve upon existing methodologies by refining databases and developing new tools that are easier to use and interpret for a broader audience of stakeholders.

Moving beyond individual process improvements, research focused on the integration of green chemistry principles across the entire supply chain is vital. This requires a systems-level approach, considering the environmental impact of raw material sourcing, transportation, manufacturing, use, and end-of-life management. Research in this area could explore the development of closed-loop systems where waste from one process becomes the feedstock for another, creating a circular economy model. Furthermore, optimizing the transport and logistics of chemicals and raw materials, potentially through the development of more efficient and sustainable transportation methods, presents substantial environmental benefits. This integrated approach necessitates collaboration between researchers, industry partners, and policymakers to encourage widespread adoption of circular economy principles throughout the entire industry.

The development of new solvent systems poses another area ripe for research. While ionic liquids offer significant advantages, research into other environmentally benign solvents like supercritical fluids, deep eutectic solvents, or water-based systems remains critical. This research must carefully consider the solvent's toxicity, flammability, and ease of recovery and recycling. Additionally, understanding the influence of solvent properties on reaction kinetics and selectivity will enable the design of more efficient and environmentally friendly processes. Researchers should also focus on developing cost-effective and scalable production methods for these solvents to enhance their industrial applicability. Addressing solvent-related issues can considerably diminish the overall environmental footprint of chemical processes.

Finally, the exploration of novel separation technologies remains a critical research avenue. While membrane separations and supercritical fluid extraction are promising, further development of energy-efficient and selective separation methods is essential. This could involve exploring advanced membrane materials or refining existing techniques to handle more complex mixtures. Improving the efficiency of these separations directly reduces energy consumption and minimizes waste generation, crucial for maximizing the sustainability of the overall process. Moreover, the development of techniques enabling the recovery of valuable byproducts from waste streams could further enhance sustainability and circularity. This research will directly translate into a more environmentally friendly chemical manufacturing industry.

7.5 Challenges & Opportunities

The transition to sustainable chemical manufacturing presents significant hurdles, demanding innovative solutions and collaborative efforts. Scaling up green technologies from laboratory settings to industrial production often faces economic constraints. The initial investment in new equipment and processes can be substantial, potentially exceeding the financial resources of smaller companies. Securing funding for research and development of sustainable technologies remains a persistent challenge, requiring both public and private investment to overcome. Furthermore, the lack of standardized metrics and guidelines for evaluating the environmental performance of chemical processes hinders the widespread adoption of sustainable practices. A common language and framework are needed to ensure consistent and comparable assessments.

Overcoming these economic obstacles requires a multifaceted approach. Incentive programs, such as tax credits and subsidies, can encourage companies to invest in sustainable technologies. Government regulations and policies can play a crucial role in driving the market towards greener options, while fostering collaboration between academia, industry, and government agencies is vital for knowledge sharing and technological advancement. Open-source platforms for sharing best practices and research findings could accelerate innovation.

Beyond economic barriers, technological limitations also pose challenges. The development of highly efficient and selective catalysts is essential for many green chemistry processes. Current technologies may not always match the performance of traditional methods, leading to slower reaction rates or lower yields. The scalability of some green technologies, particularly those involving novel solvents or reaction conditions, remains a significant hurdle, requiring further research and development to address the unique challenges of large-scale production.

Despite these difficulties, numerous opportunities exist for advancement. The development of more efficient and cost-effective green technologies is an ongoing process. Research into novel materials, such as bio-based catalysts and ionic liquids, continues to yield promising results. Moreover, advances in process intensification techniques offer significant potential for reducing waste and energy consumption. Improving the efficiency of separation processes, for example, can greatly enhance the overall sustainability of chemical production.

The growing consumer demand for environmentally friendly products presents a powerful market driver for sustainable chemical manufacturing. Companies that embrace sustainability are increasingly finding a competitive advantage, attracting environmentally conscious consumers and investors. This growing market creates an impetus for further innovation and investment in green technologies. Moreover, stricter environmental regulations are further encouraging the adoption of sustainable practices, forcing companies to reduce their environmental impact.

The incorporation of life cycle assessment (LCA) into the design and evaluation of chemical processes provides a valuable tool for identifying and addressing environmental impacts throughout the product lifecycle. By integrating LCA into decision-making processes, companies can make informed choices that minimize environmental harm while maximizing economic efficiency. This proactive approach allows for continuous improvement and optimization of processes.

Furthermore, the development of circular economy models is crucial for minimizing waste and maximizing resource utilization. Strategies like recycling, reuse, and waste-to-energy technologies offer the potential to drastically reduce the environmental footprint of chemical production. Collaboration between companies to share resources and develop closed-loop systems can enhance efficiency and reduce waste streams.

Ultimately, the transition to sustainable chemical manufacturing is a collaborative endeavor requiring the combined efforts of scientists, engineers, policymakers, and consumers. By addressing the challenges head-on and seizing the opportunities for innovation, we can build a more sustainable chemical industry that protects our planet while meeting the demands of a growing global population. The future of chemical manufacturing depends on our collective commitment to sustainability. This requires ongoing research, technological advancements, supportive policy measures, and increased awareness among all stakeholders.

www.ingramcontent.com/pod-product-compliance
Lightning Source LLC
LaVergne TN
LVHW022013200926
848481LV00007B/2367
9798348598990